Praise for

DOUGH

"Mort Zachter's small, wry memoir is . . . as miraculously loving and nonjudgmental as it is cleareyed." —*The New York Times Book Review*

"Bittersweet as a Passover chocolate bar, *Dough* provides rueful and highly readable proof of how much weirder our nearest and dearest can be than we might even dare to suspect."

—Michael Wex, bestselling author of *Born to Kvetch* and *Just Say Nu*

"[*Dough*] pays off with honest but lighthearted discoveries about loyalty and wealth." —*Publishers Weekly*

"[R]ich and entertaining . . . [I]f Zachter shares New York with Volk and Gallagher, his writerly godfather is Calvin Trillin, who wrote with affection and restraint . . . [*Dough* has a] similarity in tone—reserved and respectful . . . As is true of the best memoirists, he comes to a deeper understanding of himself—of what it means to carry on in the present, now that the past has been revealed."

—*Los Angeles Times*

"[Zachter's] memoir, with well-paced chapters that are nostalgic yet often humorous and pungent, gives a distinct take on the classic saga of working-class immigrant families struggling to succeed . . . [T]his is recommended for all libraries." —*Library Journal*

"Zachter charmingly portrays the changing Lower East Side . . . Zachter never seems bitter, describing the discovery of his uncles' secret hoard with such surpassing sweetness and affection that readers won't dream of envying his newfound wealth . . . a warm family narrative." —*Kirkus Review*

NINTH STREET
BAKERY

DOUGH

A Memoir

MORT ZACHTER

COLLINS
An Imprint of HarperCollinsPublishers

Portions of this memoir first appeared in
U.S. One, The New Jersey Lawyer, Weatherwise, Moment, Poetica,
and *Fourth Genre* (a publication of Michigan State University).

HarperCollins books may be purchased for educational, business, or sales
promotional use. For information please write:
Special Markets Department, HarperCollins Publishers,
10 East 53rd Street, New York, NY 10022.

First hardcover edition published by the University of Georgia Press 2007
First Collins paperback edition published 2008

Designed by Erin Kirk New

The Library of Congress has catalogued
the University of Georgia Press hardcover edition as follows:

Zachter, Mort, 1958–
Dough : a memoir / Mort Zachter.
p. cm. — (Association of Writers and Writing Programs
Award for Creative Nonfiction)
Includes bibliographical references.
ISBN-13: 978-0-8203-2934-5 (hardcover : alk. paper)
ISBN-10: 0-8203-2934-7 (hardcover : alk. paper)
1. Jews—New York (State)—New York—Biography. 2. Zachter, Mort, 1958–
3. Jews—New York (State)—New York—Biography. I. Title.
F128.9.J5Z27 2007
305.89'40747—dc22
[B]
2007006707

ISBN 978-0-06-166341-3 (pbk.)

08 09 10 11 12 ID/RRD 10 9 8 7 6 5 4 3 2 1

Frontspiece: Harvey Wang's storefront photo of the family bakery appeared in *Harvey Wang's New York*. The photo was also featured on a postcard for City Lore's Endangered Spaces Project.

"The life of every man is a diary in which he means to write one story and writes another. And his humblest hour is when he compares the volume as it is with what he vowed to make it."

JAMES M. BARRIE

"If we had invested in real estate, we would be rich."

HARRY WOLK

To my parents, Helen and Phil

In memory of my uncles, Joseph and Harry

For my children, Aleeza and Ari

CONTENTS

DOUGH

CASH
5.0 2

REMEMBRANCE
2006

Memory is a net: one finds it full of fish when he takes it from the brook, but a dozen miles of water have run through it without sticking.

—*Oliver Wendell Holmes*

Bread.

As a child, before I noticed much else, I smelled bread; but, had I known where to look, I would have experienced more.

In 1926, my Russian immigrant grandparents, Max and Lena Wolk, established a bakery at 350 East Ninth Street in New York City. Although my grandparents and their two sons are long gone, the business they started still exists in that very spot. It has changed, as all things do over the course of three generations. Yet, when I look at a 1960s photograph of my uncles, what I remember is that time and that place.

The black-and-white image shows two bespectacled men with short cropped hair and white-collared shirts, displaying bread and cake. Not by chance, my uncle Harry Wolk poses behind the cash register. Despite spending most of his time elsewhere—precisely where is still a point of family contention—he possessed the financial brains of the operation. Uncle Harry was always telling jokes; he was an actor on a stage, and the customers loved him. But in this photo, he doesn't even smile.

His solemn-faced brother, my uncle Joe, is more accurately portrayed in the photo. Uncle Joe never laughed, at least not that I remember, and the customers said he was crusty. A religious man, Uncle Joe moved the merchandise but would have preferred to be praying. The letters CM, HP, and P written on the boxes behind him might as well be hieroglyphics; over the years their meaning has been lost. My mom, their sister, thinks the P might stand for prune Danish, but she's not sure. She was left out of the picture.

Memory is a funny thing, more gray than black and white, and constantly evolving. To illuminate my shadows of recollection, I pieced together this montage from the surprises and stories my uncles left behind.

Let me begin this way: in their entire lives, my uncles never baked a thing.

AWAKENING

1994

I'm opposed to millionaires, but it would be a mistake to offer me the position.
—*Mark Twain*

On a sweltering August afternoon, the clatter of jackhammers blasted through the open dinette window. I sat in the hallway next to the only phone in my parents' Brooklyn tenement. Their apartment had no air conditioning—never did, never would—and my backside stuck to the vinyl seat cover of the telephone chair. The black rotary phone rang. I looked down at the dusty piece of history and imagined Alexander Graham Bell calling from the great beyond. I picked up the receiver, which felt heavier than I remembered. "Hello."

"Hi, Mr. Zachter, it's Bruce Geary."

The voice was old but quite lively. Mr. Geary sounded Irish, but I had no idea who he was.

"Yes." I was Mr. Zachter, just not the Mr. Zachter he thought he was talking to.

"There is a million dollars in the money-market account. I suggest you buy a million dollars worth of treasuries to maximize the return."

I was hearing things. No one in my family had that kind of dough. The heat had gotten to me. It must be a misunderstanding. A practical joke. I stared at the river of stains running down the walls from the ceiling. When I had lived here as a child, sleeping in the dinette with my head next to the Frigidaire, the upstairs apartment bathroom had leaked. Some things never change.

But some do.

"Hello, Mr. Zachter, are you there?"

"Yes. This is Mort Zachter. My dad is in the hospital. He had surgery for colon cancer and won't be home for a while. Who are you?"

"I'm your uncle's stockbroker, known him forty years. I've been working with your father recently."

At that moment, Uncle Harry, who had moved in with my parents two years before due to his dementia, sat in the living room, slowly sinking into an upholstered chair with broken springs, his feet resting on a well-worn patch of carpet. His uncombed hair was more yellow than gray, his face paper white; his eyeglasses rested on the tip of his nose, but his eyes were vacant. He needed a shave.

"Mr. Geary, did I understand you correctly? Did you say my uncle has a brokerage account with a million dollars in the money-market fund?"

"Yes."

I let that settle in for a minute. I didn't know how to respond. Growing up I had felt poor—not a homeless, hungry, dressed-in-rags poor, but a never-discussed sense that we simply couldn't afford better. Not better than our one-bedroom apartment, not better than vacations in Art Deco dives on Miami's Collins Avenue only in the summer, and not better than view-obstructed seats behind a pole at the old Yankee Stadium. At thirty-six, I knew lives of *not-better-than* plus a million dollars didn't add up.

Mr. Geary broke the silence. "Would you like me to send you the papers to get signed giving you power of attorney over the account?"

That was a loaded question if ever I heard one. Would I like financial control over an account with a million dollars?

"Yes, Mr. Geary, you could mail us those papers."

Mr. Geary assured me he would mail the papers out that day. I hung up and staggered to the dinette. Uncle Harry slept there now. The ancient Frigidaire that mom had defrosted by hand every few weeks when I was a child had finally been replaced by a self-defrosting GE. I perched on the edge of the bed, across from my old wood-laminate desk. A deep, jagged crack in the side of the table top ran above one of the legs. One false move and the leg would break off. When I was a child the table was whole.

Back then, the first thing I would do when I woke up was look out the dinette window. The early morning sunlight would filter through the coarse green burlap Mom had sewn for a curtain. Below our second-floor window was the cement courtyard where I had played stickball, baseball box-ball, ring-o-levio, and the other street games of Brooklyn. In the fall, the wind would rustle through the leaves of the ailanthus trees that grew in a fenced garden in the back of the courtyard, fertilized by little more than rubbish. Best known as the tree of heaven, the ailanthus thrives in poor neighborhoods; it requires only water and sunlight. I would watch the twisted blades of its seeds spiral down like airplane propellers. I loved to pick up those fallen seeds, split them open at the sticky end, and attach one to the bridge of my nose like a rhino's horn. Yet whenever I plucked one of those bright green leaves and crushed it in my hand, the rank stench it gave off had reminded me of the smell of the monkey's cage at the Prospect Park Zoo.

Mr. Geary's call left me pondering a different shade of green. No one in his right mind establishes a brokerage account and keeps a million dollars in the money-market portion of the account. Although I had never owned a brokerage account in my life, as a CPA I knew that people kept most of their investments in stocks and bonds—not in their money-market account. The whole business made no sense. Not unless Uncle Harry had more millions invested in stocks and bonds.

And that couldn't be the case. My two bachelor uncles had made my parents look rich. Uncle Harry and Uncle Joe had lived like paupers in the Mitchell-Lama housing project on Manhattan's Lower East Side. They drove a twenty-year-old Buick that had been rear-ended and looked like a giant accordion. For years, they frequented the deeply discounted dental clinic at NYU, where interning students treated them. Well into the Reagan administration, they wore suits dating back to the New Deal. Yet they owned a brokerage account with at least a million dollars in it? Could they have made millions selling day-old bread? Were they cheap, crazy, or both?

There was no one to ask. Uncle Harry could no longer tell a knife from a fork, Uncle Joe was dead, Dad was hospitalized, and Mom doled out information as if it were sugar and the world were in a diabetic coma. If I wanted answers, I would have to unearth them myself.

But I knew where to start. Earlier that day, when Mom and I had visited Dad in the hospital, he had asked me to empty Uncle Harry's post-office box.

"Why does Uncle Harry need a post-office box?" I had asked, since it was the first time I had ever heard of my uncles having one. Mom yelled at me not to ask so many questions. Dad said that Uncle Harry got a lot of mail.

Now, sitting on my old bed, I watched Mom, Brillo pad in hand, standing at the kitchen sink with her back to me. Her shoulder-length gray hair slowly swayed as she scrubbed the frying pan she had used to make Uncle Harry lunch. The kitchen had no exhaust fan, and the air was still heavy with the odor of onions from the tuna fish cakes she had deep fried. I asked her for the keys to Uncle Harry's post-office box.

She dried her hands on the white dish towel, neatly folded it, placed it on the counter adjacent to the sink, and led me to the night table next to my dad's side of their double bed. She opened the top drawer. Inside, a plastic Baggie contained Uncle Harry's keys, a black yarmulke, and the Mourner's Kaddish on a laminated card from a funeral home on Flatbush Avenue. Dad had left the bag for me in case he died. I removed only the keys as the pneumatic drilling started again.

Back in the dinette, I sat down on my old bed at the end closest to the window and slid back the faded burlap curtain. On that summer day in 1994, the courtyard was in a shambles. A demolition crew chopped away at the concrete with their drills. Ripped from its supports, the garden fence lay on its side in the dirt next to the felled ailanthus trees—my trees of heaven.

In the place where I used to play and dream and pretend, they were building a parking lot.

WAITING FOR COHN
1947

The Forecast for December 26, 1947, from the United States Weather Bureau

NEW YORK CITY AND VICINITY—Cloudy today with occasional snow ending this afternoon, followed by partial clearing, highest temperature near forty.

The city that never sleeps slept. At 3:20 a.m., on December 26, 1947, snow flurries provided a preview of coming attractions, but no one noticed. I certainly didn't. My mom hadn't met my dad yet, and I wasn't born until 1958. But I've heard Mom tell the story of that day so many times that family legend and New York City history have become one.

The snowfall officially began at 5:25 a.m. At 7:00 a.m., Uncle Joe left his footprints in the two inches of powdery snow that already covered the sidewalk in front of the Store.

My family always referred to the bakery as the Store. None of them ever called it by its legal name, The Ninth Street Bakery, or by a name I found, decades later, on some long-forgotten paperwork, Joe's Cut-Rate Bakery. The Store was a commissioned bakery; that meant nothing was baked on the premises. Instead, my uncles purchased bread and cake from wholesale bakeries in Brooklyn like Pechter's and then sold it over the counter in the Store or to restaurants in Manhattan. Customers called the bakery "the day-old bread store."

Uncle Joe removed the copper-colored Yale padlocks and pushed the front gate open. It was a powdery snow, the kind that was easy to push off your sidewalk provided there was not much accumulation. Based on the forecast, Uncle Joe figured he would shovel in the afternoon when it stopped snowing. Inside, he coerced the Store's good-for-nothing heater into action, removed his coffee from a paper bag, took the lid off, and watched the steam rise. Yesterday was Christmas. Only one bakery in the largely Ukrainian neighborhood had been open, and it had been a busy afternoon at the Store. Customers had left with cake and dinner rolls, but no deliveries had arrived.

Shortly after World War I, my grandparents had established the bakery further downtown on Allen Street. They had led a life of long hours, minuscule profits, and perishable inventory. In the early twentieth century, this had been a common way for immigrants to gain their financial independence, especially those who wanted to stay out of the sweatshops made infamous by the Triangle Shirtwaist Factory fire.

Many of the wholesale bakeries that supplied the Store used independent contractors to deliver the breads and cakes. But on that particular snowy morning in 1947, the shelves were empty, except for a few rum-and-brandy-flavored fruitcakes. Uncle Joe was untroubled. To him, fruitcakes got better with age, and customers were never a problem at the Store. The question was *merchandise*, which meant the bread and cake once it was in the Store ready for sale to customers. Who was going to deliver *stuff*, their name for the bread and cake before it arrived, on a day like this? A Friday, the day after Christmas, children off from school, the weather cold and bleak, beds snug and warm.

Uncle Joe sat on the cold white tiles of the display window shelf, sipped the sugarless black coffee he favored, stared out at the snow, and waited for the deliveries he was certain would come when the snow stopped. And it had to stop soon. In Uncle Joe's superstitious mind, snowstorms arrived in New York only after New Year's Day. He was convinced the weather was familiar with the major holidays and acted accordingly. No matter the actual weather, Uncle Joe started to wear short-sleeved shirts on Memorial Day and went back to flannel long-sleeved shirts on Labor Day.

Mom and Uncle Harry drifted in at 10:00 a.m. My mom still lived with her mother and two brothers in the house where she had grown up on Hegeman Avenue, in the East New York section of Brooklyn. After their five-block walk from the Fourteenth Street subway station, their hats and the shoulders of their coats wore a layer of snow that needed brushing off.

Ten years before, my grandfather Max had died of a heart attack in his sleep. What I know of Max comes from my mom's meager recollections. Unfortunately, like so many other people, I had no interest in family history until late in the day, when darkness had already dimmed the memories of those still alive to ask.

"Your grandfather was tall," Mom would say, though she could never say exactly how tall. "He smoked a lot. And he once sold a formula for yeast to Fleischmann's. That's how he got the money to start the Store. He had a terrific memory; he never wrote anything down."

At the time of Max's death, Uncle Joe was thirty-three and already worked in the Store full-time. But when my grandparents had immigrated to America in 1913, Uncle Joe had been nine and already steeped in the shtetl life of strict Shabbat observance and Torah study. This explains why the only place I ever saw him truly happy was in synagogue.

Uncle Harry, on the other hand, didn't know from Torah. He was a businessman. When Max died, Uncle Harry was twenty-six and still working his way through college. It would take him ten years of perseverance and a transfer from City College to the then less-demanding NYU School of Commerce to earn his accounting degree.

When her father died, my twenty-one-year-old American-born mom had already graduated from tuition-free Hunter College and was teaching elementary school full-time. Max's death diverted Mom from her chosen career. She was forced to work in the Store full-time.

My grandmother Lena died in 1961. An older cousin on my father's side of the family told me Lena was the sweetest woman he had ever met—but a lousy cook. To my mom, Lena was a capable woman: "She could do anything. She even drove a tractor in the old country." Given the time period, and their economic struggles in the Pale of Settlement (the area in czarist Russia where Jews were forced to live), the tractor must have been horse-drawn.

Outside the Store, a horse and buggy would have been most efficient on that snowy day. Seven inches of white powder had fallen in less than five hours. A car clanked as it snaked down the block, the metallic sound of its chained tires muffled by the snow.

"What do you think?" Uncle Joe said. "It can't keep up like this much longer."

"It's snowing very hard," Uncle Harry said.

"Do you think anyone will deliver today?" Mom asked.

Harry said, "Maybe Cohn—he's crazy enough to work in this weather."

"So why are we here?"

"Like I said, Helen, only the *meshuggeners.*"

According to the *New York Times,* Benjamin Parry, the U.S. Weather Bureau's chief meteorologist in New York City, noted the remarkable rate of the snowfall that morning; it was all snow with virtually no wind. So, according to him, this was not a blizzard, which, he said, was a snowstorm accompanied by high winds and zero or subzero temperatures. Lay observers called it a snow cloudburst. Descriptions did not matter. Visibility vanished. When Mom and her brothers peered through the display windows, they could barely see the buildings across the street.

At the height of the storm, between 3:00 and 4:00 p.m., the snow fell at a rate of more than three inches per hour. By then, nineteen inches of snow covered the city. The sanitation department was losing the battle to keep roads and bridges clear. "It's like trying to sweep the sea back," said one worker.

All flights were canceled at LaGuardia Field. There was no bus service in Manhattan and the Bronx. Ships entering the harbor dropped anchor, and those scheduled to leave remained at their piers. UPS hoped to resume normal deliveries the next day. Eskimos from Alaska, in New York to take part in an exhibition at the Grand Central Palace, hitched up their huskies to their dog sleds and gave them a workout on the barren wilderness of Park Avenue.

Many stores and companies closed early and sent their employees home, but getting home was a pilgrimage. After an announcer assured commuters that the trains were running on time, passengers boarded the 5:19 for Chappaqua at Grand Central Station. The announcement was correct about only the departure time. The 5:19 took seven hours to crawl from South White Plains to North White Plains—usually a seven-minute trip. The train stopped short of the station, and passengers were last seen walking through the snow on the rails. At Pennsylvania Station, passengers fortified themselves at the new liquor store; there was a bottle party in almost every car. At

6:09 p.m., service on the Long Island Railroad was suspended indefinitely.

Travelers were stranded even in Manhattan. A woman bringing a cooked turkey to a party boarded a downtown bus at Broadway and 160th Street. The bus got no further than Eighty-third Street. After several hours, she, her fellow passengers, and their driver ate the turkey.

New Jersey commuters took the subway uptown hoping to find a bus or a taxi to get them back home across the George Washington Bridge. But with no buses or cabs to be had, thousands of people, barely visible in the snowy twilight, walked across the bridge to New Jersey, collars raised, heads down, in meager defense against the storm.

By 5:00 p.m., the Store was still. Customers had stopped coming in hours ago.

"How much you think is out there now, Harry?" Mom asked.

"They said a couple of feet on the radio," Uncle Harry replied.

"What are you talking about?" Uncle Joe said. "It can't be more than a foot."

"Maybe it's time to go home?" Mom suggested.

"Wait. Let's see how it goes," replied Uncle Harry. "Getting home is no problem. We walk to Fourteenth Street, take the train to New Lots, and walk home from the station like always."

Meanwhile, groceries delivered daily—such as bread, milk, and eggs—were almost completely gone from the shelves of retail stores. There was ample supply; the problem was getting delivery trucks through. Milk trucks bound for New York City were stalled throughout Westchester. State troopers tried to escort the trucks, but even the escort cars got stuck in the snow. A man died of a heart attack on the Saw Mill River Parkway, and policemen had to carry his body for two miles because ambulances could not get through the drifts.

On WNYC, New York City fire commissioner Quayle announced, "Life and property, from fire hazards, have never been in such jeopardy." With streets practically impassable, he urged all New Yorkers to refrain from lighting their Christmas trees until the snow emergency had passed.

As the when-to-go-home debate swirled at the Store, Suzy, the bakery's black cat, lifted her head from her stained-cake-box throne on top of the heater. Then Uncle Harry, Uncle Joe, and my mom heard the muffled clank of tire chains. The faint clanking grew louder. A delivery truck came to a halt in front of the Store; the driver-side door swung open; and someone descended, disappearing into a snowbank behind a buried car. The figure reappeared, high-stepping in the direction of the Store, holding a pyramid of cake boxes aloft.

"Well, what do you know," said Uncle Joe. "It's Cohn."

Mr. Martin Cohn turned his back to the front door and pushed it in with his rear end. Inside, he turned to face his appreciative audience.

"Hi, Cowboys. I was in the neighborhood and figured you could use some bread and cake on a beautiful day like this."

It was not long before Mr. Cohn returned to his high-riding truck and disappeared behind a curtain of falling snow. The muted clanking of his tire chains was heard long after he was out of sight. His tire prints and footprints quickly filled with snow, leaving no trace of his daring. But in the Store, *merchandise* was no longer a problem.

And though thousands of cars and trucks were stuck in their tracks all over the tristate area, Mr. Cohn was not the only delivery man who completed his appointed rounds that day. At the height of the afternoon snowfall, a Railway Express employee delivered a Hartz Mountain canary to a *New York Times* pressman at his Bellmore, Long Island, home. He had ordered it from a Manhattan pet store two weeks before.

By 7:00 p.m., twenty-four inches of snow had fallen. Further uptown, Times Square, usually packed with holiday revelers, was deserted. Some bars hosted one or two commuters hoisting a few and wishing they had hotel rooms for the night. The absolute stillness was an object of wonder to the policemen on duty there. They had never experienced anything like it.

At about the same time, downtown, in Battery Park, a woman named Carly Beckwith wandered around in the storm with a twenty-four-inch-long cedar rod marked in tenths of an inch. She wore a polo coat, a bright scarf, and ten-inch-high rubber boots with lots of snow in them. She was thirty-one years old and was said to have never caught cold. But she had a dilemma. As the snow depth exceeded two feet, her measuring stick became useless. This had never happened before. She discarded her rod and started working with a yardstick, even though it was not meteorologically approved equipment. Accuracy mattered to Miss Beckwith. She was the official snowfall reader for the U.S. Weather Bureau in New York City. Her fellow Weather Bureau employee, Mr. Parry, made it clear that the only reason his department's snowfall prediction was two feet short was because the storm originated from a point off southern New England where the bureau had no observation points. Twenty-six inches of snow blanketed New York City—the deepest snowfall ever recorded in the city's history, before or since. It was even more than the Blizzard of 1888, which, Mr. Parry said, was a "real blizzard."

At about the time Carly Beckwith picked up her yardstick—the peak of the storm several hours past—customers were heading back to the Store. Mom, Uncle Harry, and Uncle Joe waited, shelves full, staring out at the snow still slowly falling. And this is where I will leave them for now, as the decade in which I was born came—and melted away—unleashing the 1960s, when I made my debut on Ninth Street.

LINDBERGH MADE IT

1994

No man was ever so much deceived by another, as by himself.—*Lord Greville*

I don't remember what I dreamed about the night after Mr. Geary's phone call. Perhaps it was the sound of pneumatic drills tearing up concrete. My nerves were so jumbled I felt as if I could shake all over, like a dog just in from the rain. What I do recall is not sleeping well.

I picked Mom up early the next morning and drove to the hospital. Dad was less animated than the day before, and his chart indicated an increase in his temperature. I figured temperature fluctuations were normal after surgery—we had been told the operation was a success—so I didn't say anything to the nurses, but I made a mental note to call Dad that night.

Surprisingly, Uncle Harry was standing in the living room when Mom and I returned to the apartment. He shuffled up to me, wearing the same wrinkled long-sleeved shirt he had worn the day before. Had I seen him on the streets of the Lower East Side, I would have thought he was a homeless person.

"Did you make the delivery?" he asked.

"What delivery, Uncle Harry?"

"You know, the delivery."

"You're not in the Store anymore," Mom screamed at him. "There are no deliveries. Stop already with the deliveries."

Bewildered, Uncle Harry listed in my direction and whispered, "Did you make the delivery?"

Mom was in too deep to understand.

"Yes, Uncle Harry, I made the delivery," I said.

"The whole order?"

"Yes, the whole order."

He started to count, either bread or money—I couldn't tell which—and in his confused thoughts, there was no difference. I told Mom I had to leave for the city to empty out the post-office box. She was surprised I was leaving so soon.

"So I can get in and out of the city before rush hour," I said.

I just wanted out of that apartment. If I'd had to deal with Uncle Harry full-time, I would have been as frustrated as Mom.

During my drive to Manhattan, I realized I had been so caught up in my own life the past few years, I had been blind to their situation. Getting fired from my job as a CPA and starting my own practice out of a spare bedroom in my Long Island home, while attending evening law school four nights a week, kept me occupied. My parents had told me about Uncle Harry's Alzheimer's and Parkinson's, but it was only after seeing him two days in a row that I realized the full extent of his deterioration.

Two years before, after Uncle Harry had undergone hernia surgery, he had not returned to his two-bedroom apartment on the Lower East Side, where an aide could have lived with him. My parents' decision to bring Uncle Harry to live with them, in new surroundings, had accelerated his dementia. Now it was too late to send him back, and my parents couldn't take care of him anymore. When Dad got out of the hospital, he would be in no condition to help Uncle Harry feed or dress himself. I dreaded trying to convince my parents they needed an aide to help them.

While circling around the block of the Peter Cooper Post Office looking for a parking spot, I remembered my two previous visits here. Only one had been during regular postal hours. The first time, I had been a kid and April 15 had coincided with Passover. We had just finished one of what I now think of as our food-stamp Seders, at my uncles' apartment, and Dad then drove Uncle Harry to this post office long after it had closed. Much to the postal worker's surprise, Uncle Harry somehow managed to stroll in through the back of the post office. How he did it, I'm still not sure. Dad and I were double-parked, and we could hear the postal workers shouting, *Hey, buddy, we're closed.* But when they saw who the intruder was—*Look, it's Harry from the bakery*—my uncle was welcomed with open arms, and his tax returns were stamped April 15, ensuring timely filing. My second visit to the Peter Cooper Post Office was in 1984, when I had picked up my passport before a trip to Israel, where I would meet my future wife, Nurit.

Back then, I worked in Manhattan, lived down the block on Eleventh Street between Second and Third, and didn't have to park to go to the post office. Not paying for parking was a badge of honor in my family. But now, after only one unsuccessful loop around the block, I gave up and pulled into a lot on Third Avenue. If Dad had known I was paying twenty bucks to park for two hours, he'd have been livid. He would have insisted I keep driving until I found a free spot on the street. And normally, I would have, but Mr. Geary's call had me curious and impatient.

It was another hot, humid afternoon in the city. After I had walked the single block from the parking lot to the post office, sweat drenched my shirt. Inside, it was silent, empty, and stuffy. I took out the piece of paper where I had scribbled the post-office box number, stuck the key into the box, opened it, and peered inside. I had never opened a post-office box before. I envisioned a hand with six fingers, like the one I used to watch as a kid on a television horror show called *Chiller Theater,* grabbing me, accusing me of opening a box that wasn't mine, and pulling me in. *Who the hell are you? This box belongs to Harry from the bakery. Where is he? Has he been filing his tax returns on a timely basis?*

No one bothered me. I was surprised my uncle's post-office box didn't have a rear wall. Through the open space, I saw postal workers sorting mail in the back. I pulled the contents out of the box and tried to look inconspicuous standing in the middle of the post office, next to a garbage can, opening a stack of mail as thick as a phone book. The first two pieces were quarterly financial statements from companies Uncle Harry owned stock in. Following Dad's instructions from that morning, I threw out these statements. Next was a monthly brokerage statement from Smith Barney. The stockbroker on the account was my new friend, Mr. Geary.

When I saw his name, I got paranoid about opening the mail in public, unsure of what I would find. I shoved all the mail into my well-worn leather briefcase. It had been given to me after a promotion at Deloitte, Haskins, and Sells, a CPA firm I used to work for. I slinked away with my paper bounty, heading east toward my uncles' apartment on First Avenue and Second Street.

On Ninth Street, right behind Cooper Union, I walked past the apartment building where Nurit and I had lived after we got married in 1985. The apartment featured both a leaky radiator and a leaky ceiling; I had felt right at home. As a fringe benefit of living in the East Village, every Friday before Shabbat, Nurit would walk down the block and purchase challah from Uncle Harry. He had charged her only the wholesale price.

When we lived in the East Village, every weekday morning,

I would put on my Brooks Brothers suit, a white button-down shirt, and a rep tie, and walk to the Astor Place station to go to work at Deloitte. Many mornings I had passed a young woman dressed head to toe in black leather, who liked to stand at the corner of St. Marks and Third sporting a live white mouse on one shoulder and a black mouse on the other. For old times' sake, I walked down St. Marks to First Avenue, past Dojo, a restaurant my uncles once supplied with bread.

Among the keys from my dad's drawer, I had found the key to the lobby entrance. The walls of the lobby were glorified cinder block, giving the building an institutional feel. It was midafternoon. I looked around to see if anyone was watching, but the lobby was empty. I opened my uncles' lobby mailbox. A torrent of envelopes exploded from the overstuffed box. I picked them up, stepped into the elevator, and hit fifteen. It took longer for the elevator to reach my uncles' floor than it took the express elevator to get to the seventy-eighth floor at the World Trade Center when I worked there for Deloitte. As the elevator crawled upward and we passed each floor, the pale light of the hallways briefly filled the elevator window. I felt as if I were traveling back in time.

Inside my uncles' apartment, darkness and dust coated everything not covered by white bedsheets. My uncles had moved here from my grandparents' house a few years after Lena died. Old cardboard boxes and newspapers were piled up everywhere. I wanted to poke around and look under the sheets, but I resisted the urge and looked for a clearing to open the envelopes.

I went through the kitchen and opened the dinette window all the way so I could breathe. Across the street, a guy in shorts was sunning himself on the deck of one of the rooftop co-ops that had been built in the 1970s. Uncle Harry used to say you had to be rich to buy one. At the dining table, I pushed the yellowing newspapers toward the other end. They fell to the floor like dominoes. I didn't bother to pick them up; they had plenty of company on the floor.

I pulled out the mail. It consisted of dozens of quarterly stock reports, a half-dozen bank statements, three brokerage statements, and a few bills. I found a July 1994 Merrill Lynch brokerage statement. The broker was Tim O'Connell, and the account balance exceeded a million dollars. The next brokerage statement I located was from Smith Barney, but the broker was Tom Koch, not Bruce Geary. This account also had a balance of over a million dollars. That meant Uncle Harry had two Smith Barney accounts. Why hadn't he consolidated the accounts? It made no sense to pay three brokerage account fees. Not unless he didn't want anyone—not even his three stockbrokers—to know how much money he had. I opened the other brokerage statement from Smith Barney with Mr. Geary listed as the broker. As I had been told, there was a million dollars in the money-market account but also quite a number of stocks and bonds.

I lined up all the July statements—from the brokers, from the banks, and one from the U.S. Treasury Department with almost a million dollars in it. That explained why Uncle Harry didn't buy treasury bills through Mr. Geary—he bought them in bulk directly from the U.S. government. I utilized my undergraduate and two postgraduate degrees in accounting and law. I did simple addition. Awestruck, I stared at the statements.

In the 1970s, I used to watch a television show about a badly injured pilot who was rebuilt as a bionic man for six million dollars. It starred Lee Majors. At the time, he was better known for his marriage to the well-endowed blond actress Farrah Fawcett than for his acting abilities. Uncle Harry was not bionic; he had never married; he didn't even own a Farrah Fawcett poster.

But Uncle Harry was the real Six Million Dollar Man.

The signs had always been there for me to see and hear, but I had never seen what was right in front of my eyes. Since my childhood, stacks of annual reports had formed meandering columns reaching ever higher in my uncles' apartment. In fact, they were still there; no one had thrown them away. I now finally understood why I remembered so distinctly my family discussing the Pennsylvania Central Railroad when it went bankrupt in 1970. They must have owned a lot of suddenly worthless stock in Penn Central. But who in his right mind accumulates the wealth of a prince yet continues to live like a pauper?

I was sure that given the chance, I wouldn't. Uncle Harry was eighty-three; my parents were both in their seventies. As a CPA, I understood estate inheritance laws and could not help but suppose that someday all this money would be mine. Expensive cars and houses did not enter my thoughts. But all the anxieties of the past few years, struggling to make a living and going to law school at night, disappeared. I felt elated; a fever broke inside me. I would never again worry about paying off my law school student loans or the two mortgages on our home. For the moment, regret over what could have or should have been remained deep inside me. But I could not understand why my parents had kept this fortune's existence from me.

Whenever my parents and I argued about my maturity, Mom would still say, as she always had, "You're young yet." I was a husband, a father, a CPA, a soon-to-be lawyer, a former adjunct tax professor at NYU, and six months past my thirty-sixth birthday. If my clients, planning their wills, asked me when children are mature enough to control a multimillion-dollar trust fund, I would always say, "This is an individual determination based upon the maturity of the person in question. But if they can't handle it by thirty-five, they'll never be able to."

Now, I went to the window and stuck my head out. The guy in shorts was gone. Only an empty lounge chair remained. The tips of my fingers blackened as I leaned on the soot-covered window ledge. The old cedar water towers, hidden in plain sight on the roofs of low-lying buildings across the Village, looked like fat rocket ships poised for take off.

"By the way . . . Lindbergh made it," Uncle Joe used to say with a smirk whenever someone made an error or asked a question that showed ignorance of what to him seemed obvious. My uncle was remembering his youth, when, in 1927, Charles Lindbergh became the first person ever to fly nonstop across the Atlantic. Lindbergh's feat had been front-page news. Everyone knew of his accomplishment.

I still remember Uncle Joe's response to the new delivery guy who brought an order of *stuff* to the Store that was missing the sponge cake: "We've had a daily sponge cake order since before you were born. By the way . . . Lindbergh made it." Or when the Second Avenue Deli called to ask where their delivery was, it made no difference to Uncle Joe that he was speaking to an important customer: "Your dozen seeded ryes? They're in transit. You'll have them for the dinner rush. By the way . . . Lindbergh made it."

And whenever he repeated his favorite saying, his eyes widened, his bushy, gray eyebrows rose and fell like windshield wipers cutting through fog, and he hesitated, just for an instant, before saying *Lindbergh*, as if that would really put you in your place.

Today it was a clear day. From my uncles' dinette window, I stared uptown at the skyline of New York—the Con Ed Building on Fourteenth, the Chrysler Building on Forty-second, and the Citicorp Building on Fifty-third. I suddenly viewed these buildings in a new way. Uncle Harry owned stock in or banked with all of them. And I would also never think of Central Park the same way. When I was a kid, Uncle Harry loved to tell me what he had claimed was Bernard Baruch's favorite joke: *What is the best investment in New York? Central Park: it opens at fifty-nine and closes at one hundred and twenty-five every day.*

Why hadn't they told me about the money? Had they thought I was immature? Or were they ashamed of accumulating millions and not sharing it with anyone? Multiple lifetimes of nothing but hard work and personal deprivation had amassed this fortune. But what good had it done anyone?

I felt the heat rising off the concrete and brick and asphalt of the city. And from a distant place, another lifetime, I heard Uncle Joe's voice—sarcastic, harsh:

Morton. You didn't know we were loaded? By the way . . . Lindbergh made it.

THE STORE
1960s

The employer generally gets the employees he deserves.—*Sir Walter Bilbey*

On Saturday, at seven in the morning, the black rotary phone in our rent-controlled Brooklyn apartment shrieked. My parents' bedroom door opened, and Mom schlepped to our telephone, located in the hallway between the bedroom, dinette, and bathroom. From my bed in the dinette, my head next to the refrigerator, I watched her plotz onto the chair next to the phone. The air whooshed out of the vinyl seat cover.

"Hello," she said to Uncle Harry.

It had to be Uncle Harry. He was the only one who called on Saturday at seven in the morning.

The dinette had no door, and I could hear his nasal voice through the phone.

"Hello, Helen. You didn't leave yet? When are you coming?"

Uncle Harry unsheathed what I had learned at an early age was the ultimate Jewish motivator—guilt. No one sat down and taught me that; no one had to. In me, it just grew.

"I have deliveries to make, and I don't want to leave Joe alone. Half the merchandise will sprout legs and walk out the door."

Uncle Harry led with the line about Uncle Joe because he knew that would get to Mom. She felt bad for Uncle Joe; if he stood all day, his phlebitis acted up. If she didn't show up, and

Uncle Harry was out, Uncle Joe, the brother she favored, never got a break and stood all day. Mom's priority was her older brother, not the merchandise.

Her voice quivered. "Soon!"

She slammed the receiver down, but not before Uncle Harry got in his usual last three words. The words that drove my mom nuts. "You belong here."

Uncle Harry and Uncle Joe worked at the Store full-time; that meant if they were awake, they were working. The Store was open seven days a week, from seven in the morning until midnight, when they went home to sleep. In 1966, they had sold their ancestral Hegeman Avenue house and moved to the First Avenue apartment six blocks from the Store to eliminate commuting time. All my uncles ever did was work. No vacations, no movies, no Broadway shows, no wives, no home life—just work.

The Store was not just a livelihood; it was their life. My uncles knew nothing else, or so I thought. Years later I learned the true reason for Uncle Harry's early morning Shabbat calls: he was alone in the Store and couldn't leave to pick up *stuff* or make deliveries until Mom showed up because Uncle Joe insisted on attending Shabbat services at the First Roumanian–American Congregation on Rivington Street, which had started an early-morning service at Uncle Joe's request.

Until I turned thirteen and prematurely graduated from Judaism, I walked alone on those Shabbat mornings to the Kingsway Jewish Center on the corner of Nostrand Avenue and Kings Highway in Brooklyn. Dad never joined me. I enjoyed services and seeing my friends. The marble cake at the kiddush after services wasn't bad either.

My mom's Shabbat mornings did not involve synagogue. Every Saturday, after Uncle Harry's call, as far back as I can remember, Mom left our apartment and walked eleven blocks to the Kings Highway subway station, took the D train to DeKalb Avenue, and changed to the double R train, which deposited her at Broadway and Eighth. Then, she walked across town for over half a mile to the Store, where she stood all day long selling bread and cake. Mom never came home before 9 p.m., and for her efforts, she received only whatever leftover bread, cake, or cookies she could carry home.

Mom not only worked every Saturday but during the week as well, filling in when her brothers had to go to the dentist or the doctor, or perhaps even take time to get a haircut. She was never paid. Family members working for free were Uncle Harry's most trusted labor source. Over the years, a rotating cast of nonrelated hangers-on did make local deliveries or pick up coffee for my uncles in exchange for bread or cake. But at the cash register, Uncle Harry trusted only family.

My memories of the Store begin with one of those Saturday evenings. When I was eight or nine, Dad and I picked Mom up and took her home. Dad didn't often pick Mom up, but when he did, it was always memorable.

Dad turned onto Ninth Street and double-parked in front of the Store, a tiny one-story shack wedged between two four-story apartment buildings. The fading autumn light shone through the adjacent apartment buildings' black fire escapes, which pushed down on each side of the Store's flat roof. Underwear flapped in the breeze on clotheslines suspended between each apartment floor overhead, making the Store appear like a sinking Chinese junk with ragged sails. Next door, a shoemaker replaced soles behind the black cat's paw sign in his storefront window. And next door to the shoe repairman, the local Democratic Club held court in a one-room office.

Above the Store's front door, a sign read "BAKERY HOME STYLE BREAD ROLLS CAKE AND COOKIES." The handwritten capital letters were a bold orange color, and a rainbow and a sun were painted at each end. The sign reflected life in the East Village in the sixties. The Grateful Dead rocked the Fillmore East around the corner on Second Avenue, and for a short time, Jimi Hendrix lived down the block. Young people packed the streets, and the air radiated with the sweet smell of pot.

"Hippies," Uncle Harry would say.

As I closed the car door, I peered through the two large display windows flanking the Store's front door to look for my mom. She was cutting a pumpernickel bread in half and didn't notice me. The orange trim around the windows peeled in strips. One window was cracked; a long piece of gray duct tape ran across it diagonally. In each window, open shelves teemed with gray cardboard cake boxes—some closed, some open, some filled, some empty—and randomly strewn loaves of bread.

Pastries and cakes also filled the shelves: chocolate-chip muffins, French crullers, jelly donuts, cinnamon Danish, marble cake, huckleberry puffs, cherry rolls, linzer tarts, almond horns, crumb buns, apple strudel, bowties (plain and sugar coated), pineapple turnovers, pecan sticks, cheese babka, coffee cake, raisin loaf, devil's food cake, dinner rings, sponge cake, and seven-layer cake. Uncle Harry told me that if I ate seven-layer cake with milk, it would be an eight-course meal. Whenever Uncle Harry told a joke I understood, which was not often, he made me laugh. Layered between, on top of, under, and inside the cake boxes were white paper bags filled with cookies. My favorites were the chocolate lace cookies Mom brought home as her pay because she knew I loved them.

I stepped into the alcove between the two front windows and

peered through the glass front door. A forest of legs grew out of a wooden floor so old that it bowed. The Store was crowded, and I couldn't open the front door to enter. Dad pushed his way in. That was how I learned the skills I needed later to enter NYC subway cars at rush hour: push first, then say *excuse me.* Dad now assumed his role as door monitor, not letting a new customer into the Store until an earlier arrival had left.

Inside was an even bigger mess than I had seen from outside. Next to Mom was a small radiator that kept you warm only if you stood within a few inches of it and leaned over. In the winter, Mom wore her coat underneath her smock. A black cat named Suzy napped on top of the heater, curled up in a grease-stained cake box. This was not the same Suzy who had greeted Mr. Cohn in 1947. Through the years, the cats would change, but my uncles always named them Suzy or, to memorialize the previous cat who had run away, Suzy-Q. Only female cats need apply for the job. If a particular Suzy found her way back to the Store pregnant, little Suzy-Q's would soon be available for the asking.

The wooden shelves behind the counter were laden with uncovered breads strewn about with no semblance of order, like a trailer park after the twister hits. Labeling the shelves was for my uncles a foreign concept. The breads included rye, pumpernickel, black bread, mixed pumpernickel/rye with its ornate swirl, and six-grain bread, which Mom said was the healthiest of all.

In her steel gray smock, Mom bellowed in a loud, annoyed voice to the never-ending line of customers, "Next!" But she smiled when she noticed me. Her straight, light brown shoulder-length hair hung limp, always parted in the middle, ruler-straight, a few strands prematurely gray. On the counter before her was a white porcelain bread scale and a large knife that she used to cut bread and cake, and, often, her fingers.

An ancient black cash register sat behind the raised center counter.

Several aging salamis hung from the ceiling behind Mom. I never understood why a bakery had salamis. I guessed that was what they ate for lunch with the black bread. Years later, Dad told me my uncles sold those salamis, and the owner of a competing bakery around the corner once accused my family of mixing milk and meat in a supposedly kosher bakery. Uncle Harry set the other bakery owner straight, challenging him to find a sign anywhere in the Store that said it was kosher. For that matter, he could have challenged him to find a sign in the Store listing prices. Prices were never posted. They varied, depending on the age of the merchandise, the time of day, the customer's looks, and Uncle Joe's mood. The only sign in the place was handwritten, in fading black marker on a piece of ripped cardboard thumbtacked to the front of a shelf. "In God we trust, all others pay cash."

Uncle Joe stooped at the other end of the counter from my mother. Salamis hung behind him as well, making symmetry behind the counter. Uncle Joe was bowlegged and looked like he was about to keel over. His hair was speckled with gray, and he wore thick, Coke-bottle-bottom glasses with dark frames. He never smiled. But when Uncle Joe squeezed a rye bread slowly and deliberately as he took it off the shelf, he reminded me of a basketball referee checking a ball for air before tossing it up at center court to start the game.

Uncle Joe rarely spoke, but when he did, he could make a few words go a long way. Speaking of Uncle Harry, who called the shots in the Store because he had the business education his older brother lacked, Uncle Joe would say, "He's the boss, and I'm the horse."

Uncle Joe said "NEXT" even louder than Mom. Other than Shabbat morning, he rarely left the Store except on those occasions when he caught the morning service at the Sixth Street synagogue. Mom referred to Uncle Joe as *Mr. Inside.*

Both ends of the counter were open, and customers walked unchallenged behind it to get a better look at the seven-layer cakes in the display window or to give a rye bread a squeeze to see if it was fresh. On the customer side of the Store, built-in shelves held metal bins heaped with uncovered bagels and bialies that customers also smelled, squeezed, sneezed on, or shoplifted, depending on their inclination. The smell of the onion bagels and bialys overpowered all competitors. My favorite corner of the Store had a bin with prepackaged Hostess cupcakes.

I walked to the far end of the counter next to Uncle Joe, where an opening with no door led to a narrow, windowless back room that was more like a long closet. When I stood in the doorway, I could smell Suzy's litter box on the floor next to the toilet. Cat urine overwhelmed the bread smell. To my left, a pile of empty cake boxes, stacked one atop another from floor to ceiling, defied both gravity and the fire code. In between these boxes to my left and the toilet facilities for man and beast to my right was a small chair and a table littered with the day's newspaper, cake crumbs, used coffee cups, and packages of saccharine. The table had a cardboard covering with names and phone numbers scribbled all over it. And although I didn't know it at the time, one number must have been Mr. Geary's. Under the table was that month's collection of umbrellas customers had forgotten and left in the Store. On a shelf above the table, a cassette player played classical music. One customer never paid with cash, and Uncle Harry had struck a deal with him to swap cassette tapes for his bread purchases.

Since Uncle Harry was out, and it was busy, Mom and Uncle Joe were both working the counter. When things slowed down, Uncle Joe would sleep in the chair in the back room, his head resting on a folded *New York Post*, then the newspaper of choice in my left-leaning family. The table also held an antique but still used bread slicer. Uncle Joe must have been very tired to sleep with his head right next to that slicing machine; it cut through bread with a sound that reminded me of a jet plane taking off.

Mom said, "Morton, sit. It will be a while until Mr. Outside comes back."

Mom called Uncle Harry *Mr. Outside* because he was always either picking up *stuff* or delivering merchandise to restaurants. At least, that was what he said he was doing.

I didn't mind staying; I loved the Store. It was an exciting place to me, packed with a choir of customers bickering over babka prices with Uncle Joe. There was surely more to see there than on the bare walls of our apartment back in Brooklyn. I sat down in the back room. I was hungry and had brought three packages of chocolate Hostess cupcakes with vanilla swirls down the middle. When I put the cupcakes on the table, the cardboard became gritty from the spilled saccharine granules. I inhaled the chalky white flour that puffed off the bottoms of the loaves Mom shuffled on the unlined shelves. Once in a while, she grabbed one and brought it to the back for slicing. To the frequent sound of the ringing of the cash register, I lost myself in the sports section of the *Post.*

Some years later, on Saturday, April 5, 1969, an article titled "East Ninth Street: A Mingling of Old World and Avant-Garde" appeared in the *New York Times.* The article discussed the stores of Ninth Street, which blended old-world Eastern European (still mostly Ukrainian) and Village bohemian. After praising the "full, rich urban life" that existed on the block, the article ended as follows:

> Not too much of this impresses the lady in the Ninth Street Bakery. She's been selling her cakes, cookies, and breads unfettered by cellophane wrapping for years. Her view of the chic universe springing up around her: *Time goes, people keep eating.*

That *lady* was my mom.

THE STORE REVISITED

1994

Is not dread of thirst when your well is full, the thirst that is unquenchable?
—*Kahlil Gibran*

I closed my uncles' dinette window and carefully returned the July statements to my briefcase, as if the pieces of paper were themselves worth millions.

I wanted to go home and share the amazing news with Nurit, but it was almost three o'clock, and I hadn't eaten lunch. Downstairs, on the west side of First Avenue, I entered a tiny restaurant and ordered a falafel and a Diet Coke from the counterman. I looked through a narrow passageway toward the back of the restaurant and noticed greenery. I walked down the low, dark-ceilinged hallway until it opened to a spacious, tree-filled backyard, where I sat down at one of the outside tables.

I felt I was no longer in Manhattan but had entered a secret garden. An oak tree provided shade, and the smell of fresh soil hung in the air. A lilac-colored rose of Sharon bush bloomed. Never judge a book by its cover, I thought.

I was the only customer, and the counterman brought my order seconds after his microwave stopped beeping. The pita bread felt warm and fresh when I picked it up; tahini sauce dripped onto my fingers. The ground chickpeas, deep fried to perfection, crunched in my mouth. I had never eaten falafel this good. I had planned to eat quickly and leave, but instead I sat and suddenly recalled another one of Uncle Harry's jokes:

A man walks into a bank, opens a suitcase, and tells the teller he wants to make a million-dollar deposit. The teller counts the cash. There is only $999,999 in the suitcase. The man insists the teller miscounted. The bank manager is summoned for a recount, and he too comes up a dollar short. The man asks for a phone. The bank manager supplies one without bothering to ask if it's a local or a long-distance call.

"Hello, Ma," the man with the suitcase whispers, "you gave me the wrong bag."

As a child, I always laughed at that joke, no matter how many times I heard it. And I heard it a lot. Uncle Harry certainly knew what he was talking about, but I was not laughing anymore. My family was certifiable. What kind of people accumulate millions and yet deny themselves the basics they could easily afford? Their stinginess made even less sense to me than why they had kept the money a secret. I felt like someone had jabbed me in the stomach with a Bayonne Flat so stale it couldn't even be salvaged for croutons. In the middle of eating the falafel, I lost my appetite.

Where did my uncles' hoarding mentality come from? To say it was just their Depression-era upbringing didn't seem right. There had to be more to it. And in later years, I learned that my uncles' hoarding ways were not unprecedented. In subsequent years, I would read in the *New York Times* about other people who echoed my uncles' behavior: a person, usually male and never married, dies alone, without children, after a lifetime working in a job one would never link to owning millions. The man might be a custodian or a clerk, someone who led a frugal life with the spending habits of a monk, perceived by others as being poor or working class at best, but someone who had the good fortune to buy stock in IBM through a dividend reinvestment plan and hold on to it for fifty years, resulting in some lucky charity being left a vast sum.

Mr. Joe Temeczko fit the mold. This family-less Polish immigrant roamed the streets of Minneapolis pushing a shopping cart filled with broken TVs and toys he fixed and sold. He dined for free at the local soup kitchen, consumed the daily newspaper at the corner candy store without purchasing it, and as a man who would have been close to my uncles' hearts had they known him, bought day-old bread. When Joe Temeczko died in October 2001, he left an estate of $1.4 million.

In our current age, when everyone, and everything, seems to have a psychological label, the term "obsessive-compulsive disorder" encompasses enough variations of hoarding to allow me to drop Uncle Harry and Uncle Joe right next to Joe Temeczko without making much of a splash. In another era, they would have been called *odd* or, if people knew they were millionaires, *eccentric*. But what proved most troubling of all was what I did not consciously recognize at that time, and what would later give me pause: I inherited my attitudes toward money from my uncles—workaholic hoarders who had no clue what charity meant. They were my example.

What I did understand, as I sat there, was my anger. No one had abandoned me or beat me or raped me. But in denying themselves, my family had deprived me as well: a bizarre flip side to the usual immigrant tradition of treating the next generation, as Alfred Kazin wrote in *A Walker in the City*, as "the sole end of their existence." And all of them did so: they all knew about the money, even my mom and my dad. And that money was as much Mom's as her brothers'. She spent a lifetime in the Store, working for no pay whenever duty, or Uncle Harry, called.

I had attended Brooklyn College, effectively a tuition-free school that charged students just a registration fee. I received an academic-based Regent's Scholarship that paid for my books. All college cost me was carfare. I lived at home. I had believed the tuition cost at a private college was beyond our financial abilities. I hadn't even bothered to take the SATs. And in college, though I had wanted to major in English, I studied accounting at my parents' insistence, so I could get a job when I graduated.

More recently, I had taken out tens of thousands of dollars in student loans to put myself through four long, hard years of evening law school, taken out a second mortgage on our home to adopt our son, worried day and night when I was fired a few years before—all while these crazies were sitting on millions. As a wedding gift, my parents had helped Nurit and me buy that leaky apartment on Ninth Street. But how could I interpret their silence in the past few years while I struggled financially?

My parents and Uncle Harry had visited Nurit and me in our Long Island home a few years before, around the time I had been fired from my job. Uncle Harry had then shuffled up to me when no one else was nearby, reached into his pocket, pulled out two shiny Eisenhower silver dollars, and placed them in my hand.

He whispered, "Morton, these are for you. Do you need money?"

At the time, I had no idea he had millions. I had stared at the frayed ends of his shirt sleeve as he held out the two coins. I took the coins but had shook my head and told him we were okay. I was not going to take a handout from Uncle Harry. I figured he would need all his money for his old age.

The commandment says, Honor your father and mother. Respect your elders. But what was here to honor? I was torn. Should I give them the benefit of the doubt? Perhaps they kept the money a secret because they feared that if I knew, I wouldn't work, would just put my feet up and wait to inherit a fortune someday? I didn't believe I would have. What I *had* inherited was the Wolk family work ethic. Lazy people don't start their own business and go to law school at night. On the other hand, what would I have done if I had known about the money? Would I have worked so hard? I knew of many cases where trust funds had produced slackers and killed great potential.

Dad would be out of the hospital soon; I had some questions for him.

I left the counterman a big tip on my way out of the falafel place and barreled up First Avenue. There was only one way to end my visit to my uncles' old haunts. I turned onto Ninth Street. Here, between First and Second Avenue, decades before, I had bicycled past the smell of shoe polish from the shoe-repair shop and the acrid odor of bad cigars emanating from the bowels of the Democratic Club's sanctum sanctorum. The bicycle came from the nearby Stuyvesant Bicycle Shop, bought for a *good price,* because the owner knew my uncles.

Since Uncle Harry had sold the Store in 1986, I had passed by only twice. In the late 1980s, it was run by an Orthodox Jewish man who kept things exactly the way they had been when my family owned it, with one exception. He closed for Shabbat. Mom told me he *ran it into the ground*; this was her way of saying he was a lousy businessman. I wondered if not being opened on Saturday had made any difference. The next time I walked by, a few years later, the Store was closed. I couldn't believe my eyes: the Store closed during normal business hours? Blizzards couldn't close the place when my family owned it. The sight of it, shut down and dark, deeply saddened me.

Now, through the late afternoon shadows, a shaft of sunlight flickered across a maroon canopy that read *Ninth Street Bakery.* Next door, the shoe-repair shop still smelled of shoe polish, but the Democratic Club had disappeared like a puff of cigar smoke. So had the Store I knew. Three-fifty East Ninth Street was unrecognizable. It had been gutted, renovated, and was clutter free. Its spotless front windows reflected sunlight without a streak of dirt. No duct tape. Inside, glass countertops sparkled. Bread and cake lay undisturbed under glass beyond the reach of customers. A black rubber industrial floor with raised circles for increased traction had replaced the old wooden floor. The ancient slicing machine was history. In front of freshly painted white walls sat a coffee pot. An NCR computerized register had replaced the ancient cash register. You could even *charge* your purchase! No litter-box smells lingered. The place was busy. But no line stretched out the door. Dad would have lost his doorkeeper job.

Instead, a young Russian woman caressed my ears with an accent as rich and sweet and familiar as a thick slice of black bread and butter. "Hello, how can I help you?"

I smiled and imagined my immigrant grandmother, with her Russian accent, standing in this very spot, moving the merchandise seventy years before. It was Friday afternoon, so I bought a challah.

What I really wanted was chocolate lace cookies. They were all out.

A NIGHT ON THE TOWN

1960s

Everybody ought to have a Lower East Side in their life.—*Irving Berlin*

Before Mom got to the back room, I smelled caraway seeds. She lifted the front of the slicing machine with one hand and placed the seeded rye inside with the other. I covered my ears with my hands and continued to read the baseball standings in the *Post*.

I loved baseball, and Dad had recently taken me to my first game in Yankee Stadium. We sat behind a pole in the last three rows of the field-level seats, so far under the overhanging lodge section that I couldn't see the fly balls. Dad told me to watch the outfielders to see where the ball would land. I never complained. Those seats were all we could afford, and I was happy just to be in the stadium.

In the Store, Mom pressed the slicing machine's handle down with her right palm to expedite the slicing. Prongs of razor-sharp stainless steel tore through the rye. When the slicer stopped, the momentary silence was shattered again as she snapped open a white paper bag with her left hand, waving it back and forth in the air in one quick motion. Holding the bag open with the thumb and forefinger of her left hand, she used her left pinky to tilt up the left side of the now-sliced rye. With her right hand, she balanced the bread upright on its end for a split second before sliding it into the open bag. She was as good as any juggler I'd ever seen on *The Ed Sullivan Show.*

"I can't leave until Mr. Outside comes back," she said. "I don't want to leave Joe alone. Go with your father; he has a delivery to make."

Dad and I loaded the backseat and trunk of our thirteen-year-old forest green 1955 Plymouth v-8 with boxes of bread and cake. Mom gave me the lighter ones to carry to the car. I was ten. We drove west down Ninth Street to Second Avenue and stopped at the light as Dad signaled for a left turn. He looked up and down Second Avenue as the blinker methodically clicked. Many buildings were boarded up and covered with posters. He turned to me in the front seat.

"This was once the heart of the Jewish theater district. All the great Yiddish actors performed here. Now they're gone."

He looked like a picture I once saw of Mahatma Gandhi: bald, with a pencil-thin mustache and a head too large for his body. Like Gandhi, Dad had trained as a lawyer. He worked as a senior claims examiner in a New York State unemployment insurance office in Brooklyn. He also worked for Uncle Harry. Every Monday night, when my dad left work, he drove to Pechter's in downtown Brooklyn and picked up *stuff* and brought it to Ninth Street. Dad was paid at the same rate as Mom. No one ever accused Uncle Harry of wage discrimination on account of sex.

Whenever we drove in Manhattan, Dad liked to tell me about the neighborhoods we were in: the flower district, the fur district, the diamond district, Times Square, and the garment district. Now, he sighed and made a left turn to head south on Second Avenue.

I could tell he missed the old days. After we passed Ratner's, we stopped at another red light. A homeless man holding a dirty rag staggered up to the late-model car stopped ahead of us. Before he could use the rag on the front windshield, the driver honked, shook his head, and waved his hand no. The homeless guy stopped, gave him the finger, and backed away from the honker, stumbling in our direction. I wondered if Dad would roll down his window and give the guy some change. Sometimes he did, but usually he didn't. Years later, when I was in my twenties and lived in the East Village, I adopted this approach as well, following my dad's example. At that moment, I felt relieved when the light changed before the homeless guy reached us. We drove past him. I stared through our rolled-up car window as he danced to music only he could hear.

Another couple of turns and we double-parked in front of a large building. More homeless people than I had ever seen in my whole life milled about in front of the building, moving in slow motion.

"There're hundreds of 'em," I said.

"This is our delivery, the Catholic Workers," Dad said. "They feed these people and help them out. Uncle Harry brings bread and cake that hasn't sold here."

"Kinda like a good deed. A *mitzvah*?"

"Let's just say Uncle Harry gives it to them wholesale."

He shut off the engine, turned on the flashers, took the car keys out of the ignition, turned to me, and said, "Wait here, I'll be right back."

He disappeared into the building. It was dark now.

Don't worry, Dad, I said to myself. I'm not going anywhere. I stretched across the front seat to his door and pushed down the lock.

In a few minutes, Dad returned with a coatless man who had followed him out of the building. He moved too fast to be a homeless guy so I figured he must be one of the Catholic workers. They started to carry the boxes into the building. The coatless man gestured for some of the homeless people to help. They crowded around the car, reached into the backseat, cleared out the boxes, and streamed into the building carrying their bounty. The stale smell of their body odor lingered in the car. Dad exchanged some papers with the coatless man, got back into the car, and turned the ignition key. "Uncle Harry should be back by now," he said.

And he was. It was late, and the Store was empty, so we didn't have to wait in line outside for someone to leave before we could fit inside. As soon as she saw us, Mom got her pocketbook from the back room and was steaming toward the front door to go home.

Uncle Harry greeted me with his usual smile.

"Morton, how are you?" he asked. "I hear you're making deliveries now."

Inside, I cringed. Ever since my classmate Janice Gellis had told me that my name meant *dead one* in French, I hated it when people called me Morton. Janice was an expert in such matters since her mother was French. I wished my parents had named me Mordechai, which is my Hebrew name. I never understood how, if I was named after my grandfather and we both had the same Hebrew name, he was Max, and I was *the dead one.*

Uncle Harry made a rare suggestion. "Helen, why don't we all get a bite at the Garden?"

Mom stopped in her tracks. "Let's go."

Mom rarely got an offer for a meal she didn't have to cook. Unlike my uncles, she had a life away from the Store. My mom loved to teach. She had gotten a master's degree in early childhood education from Hunter College, at night, and then become a substitute teacher in Brooklyn. She would have liked to teach full time, but her schedule at the Store still did not allow for it. When she started working at the Store after my grandfather died, she did so to help my aging grandmother. Now Uncle Joe was her motivator. Over the years, my mom spent a lot of time in the Store. She told me she was in the Store right after the bombing of Pearl Harbor so that Uncle Harry could finish his studies at NYU and get his accounting degree. She heard President Franklin Roosevelt's "day of infamy" speech on the radio in the Store.

Uncle Harry locked the front door so no additional customers could enter; after a few minutes, just the five of us were left. No one bothered to clean up; even if we had, you couldn't have told the difference anyway. Uncle Harry shut off all the lights and, once outside, pulled the rusty security gate across the full length of the front of the Store. It screeched like chalk on a blackboard.

We headed downtown and were soon crossing Delancey. Mom pointed out a restaurant the Store supplied called the French-Roumanian. They probably weren't kosher though, and my mom would eat only at a dairy or kosher restaurant.

On East Broadway, we passed a building with fading Hebrew writing that said *Forverts.* I read the words "Forward Building" above the arched entranceway. "What do they do here?"

"That is where they publish the leading Yiddish newspaper, the *Jewish Daily Forward,*" Dad said. "Famous writers eat next door at the Garden. That's where we're going."

I figured this Garden must be some great restaurant if famous writers went there. As an adult, I learned that I. B. Singer enjoyed stuffing himself on their rice pudding after dropping off his pieces at the *Forward.* Today, I'd be more impressed knowing that Alfred Kazin had strolled in. Even as a child, I knew I wanted to be a writer. I had already written a short story about my first trip to the dentist. I had mistakenly thought I was going to the barber. Imagine my surprise when the man with the white coat and big porcelain chair ignored my hair and told me to open my mouth.

We got lucky. Dad found a parking spot just down the block from the Garden on his first trip around the block. Inside, following Uncle Harry's lead, I grabbed a tray and stood in line behind him.

Uncle Harry was my favorite compared to dour Uncle Joe. A few years before, when my uncles still lived on Hegeman Avenue and my grandmother was still alive, she had asked Uncle Harry to plant some bushes in their backyard. I was three years old. It was the week of Passover and the Store was closed. Uncle Harry asked me to help him dig a hole in the soil with my yellow Tonka dump truck and plastic shovel. He didn't need my help, but he made me feel important because he wanted my company.

The cafeteria tray was heavy even before I put anything on it, and I liked sliding the tray back and forth on the stainless-steel tubes. A sea of vegetables—carrots, peas, spinach, corn, and string beans—steamed on the other side of the foggy glass that separated the customers from the food. I was the only kid in the Garden.

"Morton, what do you want to eat? They have cheese blintzes," Mom said.

"I don't like cheese blintzes."

"They have borscht, Morton. Do you like borscht?" asked Uncle Joe.

"Borscht, Morton, the red stuff with potatoes in it," Dad said.

"I don't like borscht, Dad. My shirt always gets dirty."

"Split pea soup, Morton? It's my favorite," asked Uncle Harry.

"Does it have cut-up frankfurters in it? I like that."

"No meat here. This is a dairy restaurant."

"I think it's a cafeteria, Dad. I'll have split pea soup and a hot roll with butter."

Uncle Harry said, "The kid knows what's good."

"Morton, what else? Soup is not enough," said Mom.

"I'll get him a piece of sole and vegetables," Dad replied before I could say anything.

At the cashier, Uncle Harry paid for himself and Uncle Joe. Dad paid for my food and Mom's. My dad was too proud to let Uncle Harry treat us to dinner, and Uncle Harry was not insisting. Even as a child, this struck me as odd. The few other times my parents had taken me out to dinner with friends—usually at The House of Pancakes—when the check came, it started a tug of war over which family would pay. With Uncle Harry, it was Dutch treat. We stood around holding our trays until Mom spotted a table for five that was not filled with old men wearing hats and loud ties as wide as they were. I didn't touch the sole. All I ate was the soup and the roll.

At first the dinner talk was about the Vietnam War and something called the domino theory, and I didn't understand what dominoes had to do with a war. But no one got really excited until they started to talk about the Store.

Mom asked, "Why can't you close one day a week, on Shabbes? Where is it written the Store has to be open seven days a week, twenty-four hours a day?"

If the Store closed on Saturday, not only would they be observing the Sabbath, but Mom would be losing her lucrative weekend job.

"Helen, Saturday is our busiest day of the week," Uncle Harry said. "How could we close on Saturday?"

"Why can't you close on Sunday?" Dad asked.

No one had an answer. Our table was silent. I heard the slurping of soup and the clatter of spoons. Finally, Uncle Joe looked up from his borscht, slowly removed the stained paper napkin tucked under his front collar, and asked, "Close shmos, Phil. How would it look if we closed any day other than Shabbes?"

Because of his religious beliefs, Uncle Joe wanted the Store closed on Shabbat. Dad was an easy target, viewed as odd man out due to his perceived lack of business sense. Dad made no reply. I sat there wondering what would happen if someone answered a question with something other than another question. Looking back, I know why—no one had an answer. There was none. The Store would be open seven days a week for another twenty years, until Uncle Joe died and Uncle Harry got too old to work anymore.

Uncle Harry smiled, waved his hand as if he were flicking away a persistent fly, and said, "It's silly, this discussion. I will work it out somehow. Did I tell you I saw Sol yesterday?"

"You were on Essex Street?" asked Uncle Joe.

Uncle Joe was jealous of Uncle Harry's gallivanting, or unsure of Uncle Harry's destinations, or both.

"Sure. On my way back from Mrs. Weinstein and some deliveries on Delancey. He tells me the pickle business is finished. Americans don't mind eating pickles loaded with preservatives that they buy at the A&P. Your grandmother and grandfather were smart, Morton. They said the bread business, that's always going to be a good business."

Dad, Mom, and I left the Garden. It was raining. It took Dad three turns of the key to start the engine and then five minutes of warming up so it wouldn't stall out; our '55 Plymouth and moisture never mixed well. Soon, we were splashing downtown on the Bowery. We passed a bank with a poster of Joe DiMaggio in front. I recognized him even though he was wearing a suit and not his Yankee uniform with the famous number five. Joe D. said we should bank at the Bowery Savings Bank. At the intersection of Bowery and Canal, at the entrance to the Manhattan Bridge, our car stalled waiting to make the left onto the bridge.

Dad couldn't get it started. A chorus of honking cars serenaded us. From the backseat, I felt the embarrassment of cold rain falling inside the car when Dad rolled down his window so a cop in a yellow raincoat and cap could ask, "Hey, bud, you got gas in this thing?"

Two cops pushed us out of the intersection to a corner spot, where we sat for a long while until our car started. This time Dad warmed her up for fifteen minutes before shifting into drive. We drove home across the bridge. As always, at mid-span Dad took his foot off the accelerator and coasted downhill to Brooklyn. The high-pitched squealing sound of the tires on the slick steel grooves of the bridge unnerved me, and I wished he would floor the accelerator to speed things up. Finally, to my relief, we hit the asphalt of Brooklyn.

On Flatbush Avenue, on our right, we passed Howard's Clothing Factory, where my bar mitzvah suit would come from; Junior's, where the cheesecake was, so Mom said, *not as good as ours and overpriced*; and a movie theater called the Fox. This made me think of what Dad had said about the Jewish theaters on Second Avenue.

"Dad, maybe they'll come back?"

"Who will come back?"

"The Yiddish actors on Second Avenue, where did they go?"
Mom answered, "Ich vaest nicsht."
That meant she didn't have a clue.

CLOSE TO THE EDGE

1994

When a deep injury is done to us, we never recover until we forgive.

—*Alan Paton*

Nurit didn't believe me until I showed her the July statements. I don't recall if we laughed or cried, most likely much of both. But there was no time either to rejoice or to feel bitter.

Later that night, I called Dad at the hospital. No answer. I called again. Someone else picked up and introduced himself. I forgot his name before he finished saying it. The doctor, who sounded very young, told me my dad couldn't talk right now but his condition was *stable.* I asked what was wrong. He told me Dad had a fever and was in considerable discomfort. I knew immediately he had no clue what was wrong with my dad. I couldn't sleep, wondering if I should go to the hospital immediately.

At six in the morning, without picking up Mom, I hurtled down an empty Belt Parkway toward the hospital. When I got to Dad's room, an empty bed and fresh sheets greeted me. He's dead, I thought. I should have come to the hospital last night. At the nurse's station, I learned Dad was still alive, but in serious condition, and had been moved to a room where *he could be better cared for.* The nurse told me the new room number, and I sprinted off before I realized I had no idea where the room was. I came back, got directions, and set off again, remembering one of Mom's favorite expressions, "If your head doesn't work right, your feet suffer."

In Dad's new, windowless room, the wall light behind his bed threw long shadows over his gray face. Tubes erupted from his arms and his nose. He whispered a weak hello to me. Shocked, I wondered how this could be happening. The operation had been a success. The nurse on duty said I should speak to a doctor. But there were no doctors with answers to be found. Many of the doctors were Orthodox, and it was Saturday morning; I walked alone down empty halls. Finally, I left, picked Mom up, and brought her to the hospital. She sat next to my dad's bed shaking her head, tears in her eyes.

My memories of the next week come back to me as a jumbled and furious nightmare. But I found a silver lining: Dr. Steve. I can't recall his last name, and believe me, I have tried to remember. He was an intern in the hospital's surgical group, and I could tell from the compassion in his voice that he cared about my dad. He wore a green lab coat. I learned later that all the doctors on the surgical team wore green lab coats. They called themselves the Green Team. By week's end, every time I saw a green coat, I turned a little green myself.

Dr. Steve could not tell me why Dad was suffering, but he did offer some reassurance and promised to speak the next day to the doctor who had operated on Dad. To remember this surgeon's name, I had to recall only a former state governor. In my mind, he would always be Dr. Governor.

The next day, I went to my dad's new room. He was gone—again. His bed was ready for a new patient. Terrified, I was now sure he was dead. I ran to the nurse's station. "Your father is in critical condition in the intensive care unit." As those words came out of the nurse's mouth, I was certain that, based on prior experience, my dad would never get out of this hospital alive.

I thought of a phone call I had received in the fall of 1975, when I still lived with my parents in Brooklyn. My paternal grandfather, Jacob, was in Beth El Hospital in East New York after collapsing in his apartment. The woman on the phone said, *Your grandfather has been moved to the intensive care unit, where he can be better cared for. Our policy is to inform the family of the change in status.* My grandfather never made it out of that hospital.

This time I asked for directions to Dad's new location before I started running through the halls. I flung open the swinging double doors to the intensive care unit. The doors were designed for patients to be rolled in alive on gurneys and, in most cases, rolled out for the funeral home. Inside, the family of one of the critical patients huddled in the hallway whispering, wearing worried faces.

Dad looked even worse than he had the day before. No one I asked—not the nurses or the doctors—had any idea why he was dying. Dad didn't have his glasses on. Without them, he was surrounded by blurs with voices. None of the nurses in the unit even knew he wore glasses. A black orderly in the unit, whose kind face grew familiar to me in the next week, saw my distress. He suggested I try his former room. I ran back but couldn't find the glasses. No one had turned them in at the nurse's station.

I left the hospital and went to get Dad a new pair of glasses. After calling his eye doctor for his current prescription, I went to a one-hour eyeglass chain store, where I chose the conservative Buddy Holly black frames my dad favored. The optometrist asked me to put the glasses on to be fitted. I told him they weren't for me but for my dad in the hospital. A look of sadness came over his face. "When he gets out, tell him to come in for a free fitting."

I slipped the glasses on my dad's face that afternoon. He said *thank you*, but he was so out of it, I was unsure he knew who he was thanking.

The next morning, when I entered Dad's room he was writhing in pain, unable to lie still. He babbled about bleeding internally and the doctors not knowing why. A nurse came in with a needle and asked me to leave his room so she could sedate him and ease his pain. Two more nurses arrived to hold Dad down while she injected him. I left his room when the nurse pulled the curtains shut around them.

I remained in the hospital the rest of the day and into the night. No one told me Dad was close to the end, but I sensed he couldn't go on much longer. I didn't know whether to stay all night, so I could be with him when he died, or to go home. Eventually he fell asleep for the night. I decided there was nothing else I could do, so I called for a cab to take me to my parents' apartment, where I helped prepare Uncle Harry for bed. After midnight, Nurit picked me up and took me home.

At five in the morning, Mom called. The hospital had called her minutes before, asking permission to operate. There were several people at the other end of the phone to verify that she gave permission. She said Dad was bleeding internally, and they were going back in to find out why and to stop it.

Dad had not been delirious the day before. He had known what was wrong with him. I headed straight for the hospital and found Mom in the lobby holding hands with Dr. Governor. Dad had survived the surgery but was still in critical condition. I asked what had caused the internal bleeding. Dr. Governor told me that when they sewed Dad up after the first surgery, they had tried to minimize how many sutures they used in "a man your father's age." They had left an opening, and that was where Dad was bleeding.

I confirmed what the doctor said. "You had to go back in because you didn't close him up correctly the first time."

Dr. Governor gave me an answer that implied he agreed. I always brought my briefcase into the hospital so I could work when Dad slept. I took out a yellow legal pad and started to write down exactly what the doctor had said. Dr. Governor stared at my writing. His face reminded me of Munch's *The Scream.* He was terrified of a malpractice suit. Mom saw his reaction.

She said, "Don't mind him. He's just a student. He just took the bar exam."

Great, Mom, I said to myself. They screw up the first surgery, almost kill dad, and you're telling him I'm just a student.

I didn't spend four long years at Brooklyn Law School for nothing; I knew the rules of evidence. In general, hearsay, an out-of-court statement offered as proof of the matter asserted, is not admissible as evidence in a court of law. In other words, if a listener hears a third party say something, that listener is not allowed to testify in court as to what that third party said. Only the third person who actually made the statement is allowed to testify in open court as to what he or she said. But there are exceptions to the hearsay rule. One of these exceptions is for admissions against interest, statements made by people showing that they committed a wrong for which they could be held accountable.

When Dr. Governor said Dad was bleeding internally because an insufficient number of sutures were used to close him, this was an admission against interest. If Dad sued the hospital, I could testify as to what the doctor had said, and that was why I wrote down his statement. I sensed Dr. Governor knew the malpractice law as well as I did. That was why he was scared shitless.

After surgery, Dad was alive but very weak and still not out of the woods. He went back to intensive care, and he was still there several days later. The internal bleeding had stopped, but Dad was still in considerable discomfort. When I entered his room a few days later, I found him sitting in a chair next to his bed, too weak to stand up unassisted. Dr. Governor had his arms around my father's waist.

"Phil," he yelled at my dad, "stand up."

Dad slumped back into the chair. Dr. Governor knew that if Dad died, I would make sure his estate sued for malpractice. The good doctor was desperate to get Dad up and out of the hospital. But to me, it looked like he was mugging Dad to get him up. I put my hand on Dr. Governor's shoulder and told him to take it easy. He lowered Dad into the chair and left the room in a huff.

On Sunday, in search of reasons why Dad was not recovering, the doctors ordered him downstairs for testing. Frustrated that it took so long to get an orderly, Dr. Steve wheeled Dad downstairs himself. Moving Dad around drained him even more, but we covered Dad with several blankets and hoped for the best. I accompanied them to a lower floor, where technicians scanned his vital organs. The results were inconsistent, and they still could not tell why my dad was sinking lower.

On Monday, I arrived at the hospital midmorning and went to Dad's room. He wasn't there. But his bed hadn't been stripped clean for a new patient. By then, I knew the routine and was not shaken. A nurse told me Dad had been taken downstairs by an orderly for testing. Outside the same room where he had been taken the previous day, I found Dad lying in the hallway on a gurney, queued up for the same scanning as the day before. I explained this to Dr. Governor and some of the Green Team who were also there. Dr. Governor turned crimson. Yesterday's order to bring Dad down for testing had not been received until this morning. When the orderly got my dad, he had not bothered to check his file to see that the test had already been done. Dr. Governor assured me this would not happen again. A flock of interns wheeled dad back upstairs. I watched them turn the corner at the end of the hall and disappear from sight. I imagined the Green Team as the medical profession's version of the Keystone Kops, with Dr. Governor doing a Buster Keaton-like cameo.

The long days in intensive care dragged on. I learned an infection was preventing Dad from recovering. One night, I came home late from the hospital, and Nurit insisted I call friends of my mother-in-law, Sylvia and Phil Lasher. Phil had had colon cancer a few years before and wanted to tell me something that would help. The Lashers had met my father several times over the years. Phil and my dad shared the same first name, were about the same age, and were both World War II veterans who had landed at Normandy. After being wounded, Phil Lasher was shipped stateside. My dad made it to Paris, decided he liked the nightlife, and somehow managed to spend the rest of the war in the City of Light.

"What can they tell me that the doctors haven't already tried?" I protested.

I wanted to sleep. Nurit nagged nonstop. I called. Like any long-married Jewish couple, Phil and Sylvia immediately started to speak simultaneously, one trying to outshout the other. I didn't understand a word.

"One at a time, please," I said.

Phil went solo and told me he had also had an infection after his colon cancer surgery. An antibiotic called vancomycin had killed the infection and saved his life. It was after midnight, but I can still recall writing down the word "vancomycin" on a scrap of paper and sticking it in my wallet.

The next morning, I asked Dr. Governor about vancomycin. Faster than it took me to pronounce it, Dad was taking it. I wondered why it hadn't been prescribed previously. I later learned that vancomycin is a powerful antibiotic and is not prescribed readily because the body may develop an immunity to it. In my dad's case, however, the potential benefit outweighed the risk.

That one word on a piece of scrap paper proved to be a lifesaver. After a few days, Dad's condition improved. I walked beside the gurney when Dad was wheeled out of the intensive care unit to a regular hospital room. The black orderly I knew from my many hours in the intensive care unit pushed a gurney through the double doors into the unit just as we were exiting. As the gurneys passed, our eyes met. He gave me a nod. He knew my Dad had beaten the odds. I smiled back, knowing how lucky Dad was to be alive.

As Dad started to recover, my conversations with my mom progressed from fearful silence to *I'm so worried. So am I,* to *What's going to be? I don't know,* and then to some of the most forthright discussions she and I ever had, before or since.

On one ride back from the hospital, Mom told me that when she was eight years old, she had won a school raffle and could choose her prize.

"My heart was set on a doll," she said. "My mother told me Harry would pick it up. The place where you got the prizes closed before she got home from the Store. Your Uncle Harry came back with a baseball bat. No doll, a bat. When my mother found out, she took me to Fifth Avenue and bought me the most beautiful doll I ever saw."

When we had this conversation, Mom had known I was aware of Uncle Harry's millions. Mom followed the doll story by saying, in effect, that when Uncle Joe died, she had expected

Uncle Harry to share some of his inheritance from Uncle Joe with her.

"But he didn't," said the woman who, for years, based on her comments, had convinced me that money meant very little to her.

After a thousand Saturdays working in the Store for no pay, even now, after the Store closed, Uncle Harry was still giving Mom the short end of the stick. Nothing had changed since my grandfather had died and my mom had to give up her teaching career to work in the Store full-time while Uncle Harry continued his college studies.

During Dad's hospital stay, Uncle Harry asked me, "Where's Phil?"

"He's in the hospital."

"When is he coming home?"

"We don't know."

With great persistence, Uncle Harry insisted on seeing my dad in the hospital. This had not been a good idea when Dad's life hung in the balance, but when his condition stabilized, there was no holding Uncle Harry back. One morning, I picked up Uncle Harry and Mom and took them to the hospital. When Uncle Harry saw Dad, he shuffled to his bed, leaned over, kissed my dad on the cheek, and hugged him.

"Phil, how do you feel?" Uncle Harry asked.

Mom and I looked at each other in disbelief. Uncle Harry had not been that cogent in weeks. Dad said he was fine. My eyes watered. Until that moment, I had never realized how much Uncle Harry loved my dad. Six months later, when I found the real-estate closing papers for the Hegeman Avenue house that my dad, the lawyer who didn't practice, had completed for my uncles in 1966, and the sales contract Dad had drafted for Uncle Harry in 1986 when he sold the Store (for the grand total of twelve thousand dollars), those papers confirmed how close the two men were and how much Uncle Harry trusted him. Dad knew all the facts of Uncle Harry's financial life and had for years.

A week after Dad escaped intensive care, the hospital discharged him. The morning I took him home from the hospital, we walked out the main entrance and across the street to the hospital parking lot. A fee-paying parking lot.

"Why did you park here?" Dad asked.

"Tough to park around here. And you also almost died a few days ago. This way, you'll walk less."

Dad didn't argue. But I knew he felt I should've found a free spot on the street. Same old stubborn pain in the ass, I thought. And that's when I knew he would recover.

I remained furious at Dr. Governor. Had Dad died, I would have sued for malpractice. But Dr. Governor had not intended injury; he had just tried to reduce an elderly man's discomfort by using fewer sutures. As Dad got stronger, I came to understand this. My anger subsided. Dad was my example. He was aware a wrong had been done him, but he also knew there was no bad intent. He let it go. He could have sued but didn't. If a man who had suffered so intensely could put it behind him, so could I.

THE FOOD-STAMP SEDERS

1960s

One must make changes on this night, so the children will notice and ask, why is this night different?—*Maimonides*

For one week each year, beginning on the first night of Passover, the Store closed. No self-respecting Jew could sell bread that week. The Store closed, but we never stopped working.

On the afternoon of the first night of Passover, Mom telephoned the Store to ask Uncle Harry, "When will you be up in the apartment?"

Passover started at sunset, and she needed to get into my uncles' apartment to start preparing the festive meal, but she didn't have a set of keys.

"We are closing up right now," Uncle Harry said. "We'll be there in five minutes."

After lunch, Mom, Dad, and I drove from Brooklyn to my uncles' East Village apartment. The trip usually took an hour, depending on traffic. When we got to the apartment, no one was there. We drove to the Store. Uncle Harry and Uncle Joe were still serving customers. Uncle Harry gave us bags of dinner rolls to deliver to a rock club called Max's Kansas City. He also gave us the keys to the apartment. From Ninth Street, Dad drove uptown to Park Avenue South. It was after 3 p.m. Mom stewed in the passenger seat next to Dad, complaining how she wouldn't have enough time to finish making dinner before the holiday started.

It was a brisk afternoon, and I struggled carrying a tall bag of rolls into a dark, narrow room with a long wooden bar. Max's, a forerunner to CBGB's, didn't last.

After making the delivery, we circled lower First Avenue until we found a free parking spot on the street. Inside my uncles' apartment building, the elevator door's small diamond-shaped window was cracked. Upstairs in the fifteenth-floor hallway, Dad and I held the bags of groceries we had brought with us from Brooklyn and waited outside the apartment door. Uncle Harry had so many keys on his key ring, Mom couldn't figure out which were the right three keys. What were all those keys for? After five or ten minutes that felt like hours, she got all three locks open at the same time, and we entered. Mom dashed into the kitchen and put on her smock. Within minutes my grandmother's white kosher-for-Passover porcelain meat pots with the thin blue rims covered all four burners. Soon the kitchen smelled of flanken, a meat my uncles liked because it was soft and easy to chew. As usual, we would sleep over on the first two nights of Passover—Dad and I in the single beds in the second bedroom, Mom on the sleeper sofa in the living room. Dad unpacked the little red-plaid cloth suitcase he and I shared.

With great difficulty, I forced open the door to the apartment's terrace and stepped outside. Only a small plant stand with a green plastic pot filled with dried soil was on the terrace.

"Don't get dirty out there, Morton. Your uncles never use that terrace," Mom yelled to me from the kitchen. "It must be filthy!"

The view was magnificent. I looked west across Manhattan over the low-lying buildings of the Village and uptown to the midtown skyscrapers. It wasn't until a couple of hours after I had watched the sun set that Uncle Harry and Uncle Joe showed up.

Forewarned by the jingle of Uncle Joe's keys, Mom greeted Uncle Harry at the door with a less than sisterly, "*Good morning,* I'm glad you could make it. Would you like to go back and make some more deliveries? Maybe Dojo needs more stuff?"

By now, Mom had finished her preparations. In the dinette, a white tablecloth she referred to in Yiddish as a *tishtoch* covered the rectangular Formica table with chrome edging. The green set of Depression-era Passover meat plates were out for their annual appearance. In accordance with the laws of kashruth (keeping kosher), one set of meat dishes and another for dairy were used only during Passover week. Uncle Harry had changed into a clean white short-sleeved shirt and was sitting at the head of the table facing the open window, his back to the kitchen and my mom. Mom never sat; she just served.

In front of my place setting was a Manischewitz Haggadah. I never thought there was any other. A Haggadah is the story of the Jewish exodus from slavery in Egypt, and as I learned only years later at Seders with Nurit's family, there are literally hundreds of versions, some incredibly beautiful. But you had to buy those. Manischewitz, a kosher wine manufacturer, gave them out as a free promotional item featuring pictures of bottles of their wine on the cover. Mine was the only Haggadah at the table.

According to tradition, the youngest child sings the Four Questions. Since I was twelve and the only child present, the job was mine. I did not need the Haggadah. I knew the whole thing by heart. Mom stopped scurrying around the kitchen as I sang the Four Questions in Hebrew, which begin:

Ma nish-ta-na ha-lai-la ha-zeh, mee-kol ha lei lot?
How is this night different from all other nights?

I finished to a chorus of "Great, good job, wonderful."

But there was no answer to the question, "Why is this night different from all other nights?" No one said kiddush, the blessing on the wine, before my chanting of the Four Questions. Afterward, no one broke the middle matzo for the hiding of the *afikoman*. I did not know what an *afikoman* was. After I was married, I learned that children at the Seder hide the *afikoman* and return it at the end of the Seder in exchange for a gift or monetary reward. According to this custom, the Seder cannot end until the *afikoman* is returned. But in my uncles' apartment, the Seder never began since we never read or discussed the Haggadah. The only sound was the slapping of Mom's house slippers on the linoleum floor as she brought in the soup plates for the first course of chicken soup and matzo balls.

The main course of meat and potatoes quickly followed. Boxes of Streit's matzo sitting on the table were soon empty, washed down with Fresca. No one drank alcohol, not even the four required cups of wine. At this point, either Dad or Uncle Joe, I can't recall whom, realized Mom hadn't eaten anything.

A cry went out. "Helen, sit down and eat."

I brought a chair in from the living room and placed it next to mine. Mom put together a plate for herself and sat down.

Dessert was kosher-for-Passover cake, of which there was always more than enough to go around.

Uncle Harry now turned on the radio on the kitchen counter.

With News Radio 88 in the background, he said, "It's time to work."

Mom cleared the table of all plates and removed the *tishtoch.* Uncle Harry dumped a year's worth of food stamps onto the table. The food stamps had been collected at the Store and had remained uncounted since last Passover. They had to be submitted for conversion to legal tender by a certain date or they were worthless. Passover Seders were an opportunity for Uncle Harry to have extra sets of hands counting, stacking, and putting rubber bands around ever-growing piles of food stamps. We counted. We talked.

As usual, the discussion came around to the Professor. Uncle Harry rubber-banded a stack of food stamps and said, "If we had only listened to the Professor."

Mom turned to me and explained, as if I had not heard the same explanation the year before, "You see, the Professor was the head real-estate appraiser in Manhattan for many years, the top man, and he was a regular customer, and he just loved your Uncle Harry."

At which point, Dad replied, "And any time a good building was for sale in bankruptcy proceedings in the neighborhood, the Professor would tell your Uncle Harry about it and tell him he should buy it."

"But we never bought," Uncle Joe added quickly, shaking his head and rolling up the sleeves of his brown-plaid flannel shirt.

The discussion always ended with Uncle Harry saying, "If we had invested in real estate, we would be rich."

News Radio 88 droned on late into the evening, repeating the same news every half-hour. From the open dinette window, the music of the East Village night rose from First Avenue: fire-engine sirens, roaring motorcycles from the New York City headquarters of the Hell's Angels on Second Street right down the block, the occasional scream, and, as trucks passed over them, dull reverberations from the heavy steel construction plates that covered pothole-filled lower First Avenue. It was past midnight when we called it quits. I assumed we had to leave some food stamps over for the second Seder the next night.

I looked forward to it.

MONEY TALKS

1994

With money in your pocket, you are wise, and you are handsome, and you sing well too.—*Yiddish saying*

I practiced triage when Dad was in the hospital, dealing with the most critical patient first, and then tending to the walking wounded. My hospital struggles took priority over caring for Uncle Harry.

I wanted to hire someone to come to the apartment to help with Uncle Harry. With only my parents to care for him, he didn't even get outside daily. One of the basic precepts of elder care is to keep Alzheimer's patients in their natural surroundings as long as possible by using caregivers who come to the home. The continuity is a comfort to someone with memory problems. After two years in my parents' apartment, for better or for worse, it was Uncle Harry's current home. I couldn't get him back to his Manhattan apartment, but at least I could get him the right help in Brooklyn. Through the hospital's social-services office, I was referred to an agency that placed in-home caregivers.

With my dad still in the hospital, I hired an aide to come to my parents' apartment every day. The morning this service began, I made sure I was in the apartment before the scheduled start time. When the buzzer downstairs rang, I ran down and opened the lobby door for a gentle-looking woman who reminded me of the actress Esther Rolle: she was heavyset, with short-cropped hair and a smile that gave me hope this plan would work out. She listened intently while I explained that my uncle had Alzheimer's and Parkinson's and needed help eating, showering, getting dressed, and taking a walk outside every day. Her warm eyes said she understood. She told me she had just been working with an Alzheimer's patient on her last job.

"Perfect," I said.

I brought her upstairs, introduced her to Mom and Uncle Harry, and left for the hospital. On my return later that day, both Uncle Harry and the bathroom were scrubbed and clean. Though not pleased with the arrangement, Mom knew she needed help. For the next week, things went well on the apartment front.

Then Dad came home, and the shit hit the fan. I received unsolicited, daily commentary on the issue of visiting help. "What do we need it for? What does she do?"

Which Dad answered himself. "She just sits all day."

I knew this was not true. My parents' apartment was cleaner than it had been in years. You don't clean a bathroom as filthy as my parents' by sitting around all day. Uncle Harry got outside daily. I didn't dare mention to my parents that they now had someone to look out for Uncle Harry so they could do something fun for themselves, like attend a Wednesday matinee. My dad's whining was not intended to start a conversation. I wasn't going to argue with him. I had been there before.

When I called the agency to get this wonderful woman back for the next week, the agency owner told me my parents had resisted help coming to the house. It sounded like this was a problem he had run into before.

I said, "I promise she will be paid for the next week."

The voice of experience answered, "If they don't let her in one day, that's fine. You will be billed for the day. But if they do that and tell her not to come back anymore, then she has no work for the rest of the week. We book weekly, and she won't get paid by us for the balance of the week. That's our policy."

I naively replied, "Don't worry about it."

On Monday, Dad didn't let her into the apartment when she arrived for work. I felt like a horse's ass when I called the agency owner and had to listen to his I-told-you-so speech.

My hands were tied; I was out of choices. The week before his colon cancer surgery, my parents had taken Uncle Harry to a nursing home and had him admitted one Sunday.

Dad said, "We couldn't leave him there. They took him to a big room where all the patients were together. It was noisy and crazy there. We just couldn't leave him."

They didn't. Instead, they took him back to the apartment and left him alone while Mom ran to the hospital. When Dad explained their attempt to place Uncle Harry in the nursing home, I understood. My parents couldn't put Uncle Harry into a nursing home: they were too close; they couldn't do it to him. This was now my job.

Through referrals from the hospital, I learned that the two best Jewish nursing homes in New York City were the Sephardic Home on Cropsey Avenue in Brooklyn and the Jewish Home for the Aged in Riverdale. Riverdale was a schlep, but the Sephardic was near both my parents' apartment and my home in Woodmere. When I called the Sephardic Home, however, I learned they had no immediate openings, and there was a waiting list. Admission could be months away. The situation in the apartment couldn't remain as it was that long. Two years of taking care of Uncle Harry had drained my mom's patience, and my recuperating Dad only made the job harder.

As they had before when the chips were down, Nurit's family helped. Her uncle and aunt heard about the long waiting list and asked if I wanted them to try to speed up the process. I gave them the go-ahead. They contacted a rabbi they knew who was somehow affiliated with the Sephardic nursing home. Before I knew it, I had an appointment to visit the home to meet the assistant director.

Nurit and my mother-in-law, Ellen, came with me on my visit. Years before, Ellen had worked in a nursing home as an activities director. Nurit had visited her mother at work many times, so she too had experience judging the quality of a nursing home. The assistant director took us to the section of the home for Alzheimer's patients in the back of the building, accessible to an enclosed outside area. Several patients sat around watching television. We walked through the clean and well-maintained cafeteria. The patients were not all in one big "loud and crazy" room. We all agreed: as nursing homes went, this one had good facilities and employees who cared about the patients.

After the assistant director completed our tour, I met with the assistant financial person for the paperwork.

"Mr. Zachter, will your uncle be on Medicaid or private pay?" she asked.

A simple question, but the answer had important ethical implications and would also determine when Uncle Harry would be admitted. If I said Medicaid, this meant my uncle's assets were below certain statutory minimum levels that required the nursing home costs to be paid by New York State. In that case, the nursing home's daily compensation would be state-mandated and less than the amount they would receive if Uncle Harry was a private-pay client. A private-pay patient meant that the patient, or his family, would be paying the nursing home two hundred dollars a day out of their own pockets.

At that time, nursing homes preferred private-pay clients in order to maximize their daily compensation rate. However, many patients were on Medicaid because their assets were below the minimum requirement, or because they had transferred the applicant's assets from his or her own name into a trust to a relative. If this transfer was completed a legally mandated number of months before the Medicaid application was filed, it was perfectly legal. The statute had been created to allow people who had struggled all their lives to accumulate a modest nest egg not to have it eaten up by nursing home costs. But many applicants transferred huge sums out of their own names early enough to qualify for Medicaid. A millionaire, with a good attorney who specializes in Medicaid planning, could shift the entire cost of his or her nursing home stay to New York State. A whole industry called elder care had developed for just this reason.

Not my uncle. Even if I had known Uncle Harry was a millionaire early enough to do this type of planning, I wouldn't have done it. This type of planning was legal for wealthy clients like my uncle, but to me it was an unethical end-run around the Medicaid statute.

"Private pay," I said.

"Does your uncle have significant assets?"

Lady, you haven't seen significant until you get a look at his assets, I thought to myself.

"Yes."

"Good. Please fill out his financial data along with the application."

"No problem."

"You understand, Mr. Zachter, as part of the application process, you will have to substantiate your uncle's ability to pay?"

At over two hundred dollars a day, a year in the Sephardic nursing home would then cost seventy to eighty thousand dollars. The nursing home needed proof Uncle Harry could afford a sum that large.

"I'll send you the substantiation you need."

That afternoon I mailed her copies of brokerage statements with the names Geary, Koch, and O'Connor at the top, and Colgate, Exxon, and Pfizer at the bottom. I didn't have to wait long. I got a phone call two days later asking me to bring Uncle Harry to the nursing home for a physical.

The "physical" was performed by a doctor employed by the nursing home, but it was really an interview. The nursing home evaluated each resident to make sure he or she would not be violent or disruptive.

The next day, I was nervous as I drove to pick up Uncle Harry and take him for his interview. What if he didn't want to go? What would I do then? Tell him we were going for a delivery? My parents had explained to him that I would be coming to take him to meet some doctors at a nursing home where he might go to stay. I was glad I had not been there for that discussion. Uncle Harry was calm when Mom and I drove with him to the nursing home. Whatever they said had worked. Dad stayed home. He said he was not up to going. I understood.

We sat in the airy reception lobby waiting for someone to bring us upstairs. I overheard two elderly female residents talking to each other. They were quite loud, and I couldn't help but hear them. I recall only one sentence from their conversation, but I remember it word for word: "No matter what shape we're in, we all fight to stay alive just one more day."

A rush of sadness hit me. I was placing Uncle Harry in a nursing home. It was a very good one, but this was his final destination. He was going to die here. Uncle Harry sat next to me. Though I wondered what he was thinking about all this, I didn't ask him.

The tiny examination room was cold and antiseptic, but the interviewing doctor had a beard and a sense of humor. He even made a few jokes. But Uncle Harry was not laughing. The doctor started with a series of questions testing memory, or lack thereof.

"What is your name? What day is it? What month is it? What season is it? What year is it?"

Uncle Harry knew his name, but he could not answer any of the other questions.

"Mr. Wolk, can you tell me the president's name?"

His eyes closed, and his face scrunched up.

"The actor?"

"You mean Ronald Reagan?" the doctor asked back.

"Yes."

We all laughed, and that broke the ice.

Although I think he joined in just to be part of the crowd, even Uncle Harry laughed. Bill Clinton was then president, but it did not matter. I watched the doctor's face while he laughed and decided my uncle had the right stuff to be admitted to the Sephardic nursing home. He was no threat to anyone. In his Alzheimer's-induced dementia, violence was not his problem.

Mom and I waited outside the examination room as they gave him a real physical. That took less time than the interview. Soon, Uncle Harry and Mom were back downstairs in the lobby waiting for me to finalize the paperwork.

I again found myself seated across the desk from the nursing home's assistant financial person. The file of Harry Wolk was open on her desk; I saw the copies of the brokerage statements I had mailed her on top. She asked me how my uncle had accumulated such a vast sum of money. I told her my family owned a retail bakery on the Lower East Side for over seventy years. Her raised eyebrows indicated doubt.

What was I going to tell her? That Depression-era workaholics who deny themselves basic wants can amass millions, or that Uncle Harry was a selfish old scrooge who got filthy rich screwing over those closest to him and never shared his millions with anyone, or that my family suffered from a collective anxiety-based obsessive-compulsive disorder that manifested itself in hoarding money.

I just said, "They also sold wholesale."

She nodded her head yes. This response, she believed. I tried not to laugh. She smiled. Her tone warmed, and I had the distinct sense I had a new best friend. She assured me the Sephardic nursing home was a wonderful place where people "as successful as your uncle" were well taken care of. This was the first time I was on the receiving end of the assumption that having great wealth automatically made you a good, and deserving, person. Or did she say this to kiss ass, and not really believe it?

"Yes," I replied weakly. Right or wrong, Uncle Harry and his dough were at the top of the list for entry to his final destination.

As we drove home, I thought about what that meant. This was not like waiting for a table to free up at the Garden. We were waiting for someone to die so a bed would open up. Less than a week later, I received a call from the nursing home. A semi-private room was now available on an upper floor. Mr. Harry

Rosenblatt had died at the age of 101. The next day's *New York Times* obituary said he was "one of the last survivors of the Jewish Legion of World War I, which fought with the British against the Turks in Palestine." Taps for Harry Rosenblatt meant reveille for Uncle Harry on Cropsey Avenue. Or was it reveille for Harry Rosenblatt, and taps for Uncle Harry?

One image sticks in my mind from the day we deposited Uncle Harry in the Sephardic Home. As I walked into my parents' apartment to drive Uncle Harry to the nursing home, I saw a small red-plaid cloth suitcase, packed and ready, sitting upright in the hallway for one final journey. It was the same suitcase we used when we slept over at my uncle's apartment in Manhattan for Passover and Rosh Hashanah. The next day was Rosh Hashanah.

My parents came along to bring Uncle Harry to the nursing home. *They* were no longer placing him in a nursing home. I had done the dirty work. I had signed the papers and made the calls. I did what their consciences wouldn't allow them to do. Now, they were just along for the ride—to ease Uncle Harry's transition into his final destination. I was behind the wheel on this delivery.

No one said a word during the car ride. At the nursing home, we took Uncle Harry up to his room and met his roommate. Uncle Harry's roommate was still with-it mentally, and he told me about his hobbies and how he kept himself busy. I liked the guy. Mom didn't care for him because his bureau infringed slightly on Uncle Harry's side of the room.

The room was furnished in mirror images—two bureaus and two chairs, one in front of each bed. Like a magician pulling a rabbit out of a hat, from her oversized pocketbook, my mom pulled out a framed photograph of a handsome, young

Uncle Harry when he graduated from NYU in the 1940s. I had never seen the photo before. She gently placed it on top of his bureau.

"Let them all see what he used to be like," she said. Her voice blended sorrow and anger.

I went downstairs for some last minute paperwork while my parents checked out who Uncle Harry's aide was, where Uncle Harry would eat, what he would eat, and who would feed him.

Downstairs, someone in admissions asked me if I wanted to leave fifty dollars as a cash draw if Uncle Harry wanted to buy something in the general store. I told them I didn't think my uncle would be buying anything; he could no longer handle money because of his Alzheimer's. They said they understood. I left the fifty dollars anyway. Uncle Harry never made it to the general store.

Upstairs, my parents were still fussing over Uncle Harry as he stood at the doorway to his room. Uncle Harry's roommate was nowhere to be seen, but I did notice that his bureau was now pushed to his side of the room.

Uncle Harry started to walk down the hall in the direction of the staircase, begging more than asking, "How do I get out of here?"

When he had asked me the same question eight years earlier in Beth Israel Hospital, where he was convalescing after a minor stroke, I had known he would eventually go home. Now he was home. We cut him off before he got to the stairs. None of us had an answer for him. A no man's land of silence stretched between Uncle Harry and the three of us, as we stood shoulder to shoulder blocking the stairs. At that moment, a man of about seventy who volunteered his time at the nursing home after his wife had died there came to our aid. He introduced himself to

Uncle Harry. The volunteer told us services would be held at the nursing home for the Jewish High Holidays the next morning.

"Would your brother like to attend services?" he asked Mom.

"Of course."

He told Mom he would bring Uncle Harry to the synagogue downstairs tomorrow for Rosh Hashanah services. This gave Mom a chance to fuss about the nature of the services they conducted. She made it clear Uncle Harry was used to an Ashkenazi, or Eastern European, service, not a Sephardic, or Middle Eastern, one. *Come on, Mom,* I said to myself, *give the poor guy a break.* The volunteer had calmed things down and gotten Uncle Harry focused on something other than a breakout. Uncle Harry had no clue about the finer points of praying anymore; stylistic differences in a High Holiday service were beyond him. If you had taken him to a mosque, he wouldn't have noticed.

My parents told me they would stay there the rest of the day. I should go home to my family; they would take the bus back to their apartment. I said good-bye to Uncle Harry and told him I would visit on Sundays. Leaving him in that nursing home was even harder than bringing him there. I rang the elevator button and waited. Down the hall, Uncle Harry was again asking, "How do I get out of here?" That night I called my parents to see how things had gone after I left. They said Uncle Harry had kept trying to leave the floor, so the aides put him in a restraint. I didn't ask for details.

I didn't cry on my car ride home. Uncle Harry's millions had finally afforded him a room on a high floor, with ocean views and three cooked meals a day. But it was his final destination—a nursing home.

THE MILLIONAIRE

1960s–1970s

I am not rich. I am a poor man with money, which is not the same thing.

—*Gabriel García Márquez*

Uncle Harry, Uncle Joe, Mom, Dad, and I sat around the Formica table in my uncles' apartment. It was Rosh Hashanah.

In my mind's eye, I can still see the dirty dinner dishes piled up in the sink waiting for Mom to wash them. A low flame blackened my grandmother's soup pot. The smell of chicken soup permeated the dinette. Brown crumbs of *wasser challah* crust dotted the white tablecloth. Every once in a while one of us leaned forward, took a sip of Lipton tea, and noshed on a piece of *kichel*. The slurping of the tea, the clinking of the teacups back onto the saucers (it was a holiday so we were fancy and did not use glasses), and the crunching of the hard, sugar-coated dessert biscuits formed a rhythm.

"Can you believe it?" Uncle Harry asked of no one in particular.

"Penn Central," Uncle Joe replied.

"The interstate did them in," Dad concluded.

Mom said, "Imagine a company like that going under."

"Penn Central," Uncle Harry whispered and shook his head. His wavy hair fluttered about. He needed a haircut desperately but said he never had time to get one.

I was still a kid and had no idea what the fuss was about. Why did they make such a big deal about a railroad that had gone out of business? No one in my family used the railroad to get to work. It never occurred to me to ask why the Penn Central Railroad was important to them. It may have been Rosh Hashanah, but for me it was Passover all year round, and I was the son who did not know how to ask. Traditionally, during those Passover Seders, when the answers to the Four Questions are actually read, the discussion centers around how four hypothetical sons—one wise, one foolish, one evil, and one who does not even know how to ask a question—respond to the proceedings. My uncles, Mom, and Dad all knew the answer to the question I never knew how to ask. But with this crowd, asking questions when you really wanted to find something out was not encouraged. I didn't have any brothers or sisters, wise or foolish, so things got real quiet real fast.

Uncle Harry broke the silence.

"We need to raise prices in the Store."

Uncle Harry had more on his mind than raising prices. As he would suggest many times over the years, he wanted me to work at the Store. He spoke of how busy they were and how they needed a hand. Uncle Harry tested the waters.

"Helen, why don't you bring Morton one Saturday? We'll show him the business."

Mom screamed, "No! Never! He should waste his life running to deliver to crumbs like Dojo, so their tabs can grow even higher? I don't want him ending up like the two of you—married to the Store."

Mom paused to catch her breath. "It's time you put a lock on the door once and for all." That was her subtle way of saying it was time for my uncles to retire.

For the moment, Uncle Harry didn't bother to reply. He knew he was licked. He lifted his hands, palms up, in hopes of avoiding another blast. Mom raised her head majestically at a slight angle. At that moment, despite the chicken-stained smock drawn tight around her middle, she was truly regal.

She looked directly at Uncle Harry and announced with great pride, "My Morton doesn't like money."

I never knew what she meant by that phrase. But by the way she said it, I was convinced they gave out awards for not liking money. If I worked in the Store, Uncle Harry was not going to pay me in cash. Cake maybe, but not cash. So I figured there was no way I could get rich working at the Store. That couldn't be what she meant.

I knew my uncles were poor. A few years later, Uncle Harry would come to my bar mitzvah at the Kingsway Jewish Center in Brooklyn solo. I assumed my uncles couldn't afford to close the Store on a Saturday and lose the revenue from their busiest day of the week. Devout Uncle Joe manned the fort, missing my spiritual high-water mark. Uncle Harry ran out right after the ceremony anyway, telling me he couldn't stay for the luncheon since he had to get back to the Store to relieve Uncle Joe.

I understood the phrase I *didn't like money* to mean that I didn't need much. Wants were out of the question. Long after that Rosh Hashanah, I heard I did not like money so many times I began to believe it. Yet, as a child, surrounded by the barren, pictureless mint-green walls of my parents' Brooklyn apartment, I spent many hours camped out in front of our black-and-white nineteen-inch Zenith mesmerized by reruns of a television show called *The Millionaire.*

I watched *The Millionaire* alone. It never interested my parents; they preferred *The Odd Couple.* I shut off all the lights in our living room, drew the two window shades down all the way, pulled out the plastic ON button, and watched the picture tube come to life. It started as a white circle the size of a pinhead surrounded by black. I moved the antenna on top of the television back and forth until the reception came in with the least amount of snow. Then I lay down on the carpet, leaned my head against the cloth sofa encased in clear heavy plastic, opened a bag of chocolate-lace cookies my mom had brought home from the Store, and began to eat. Mom was in the kitchen. Dad sat in the foyer smoking Kents and reading the *New York Post.* I heard him strike a match and smelled his smoke in the living room as the show started.

"Hi, my name is Michael Anthony. For many years, I was executive secretary to John Beresford Tipton, financial genius, philanthropist, and billionaire."

Gates opened onto a tree-filled estate with a huge mansion.

"And this is Silverstone, where he retired and began the hobby of giving away to a randomly selected man or woman an amount in the sum of one million dollars, tax free, as a gift with no stipulations or requirements attached."

Inside the mansion, John Beresford Tipton sat in a huge chair with his back to the camera. Viewers never saw his face. I imagined Nelson Rockefeller. Who else would a billionaire look like? Michael Anthony entered to pick up the envelope with the name of that week's new millionaire.

In one episode a single woman lived with her ten-year-old son. She worked in a department store. Her boss was a jerk. She wanted to quit but couldn't because she needed the money.

Then Michael Anthony knocked on her door and gave her the million-dollar check. The nice fellow she was dating was not rich. He was just your average guy with a job. She believed that if she kept the money, her relationship would end. A woman with more money than her husband was not possible in this 1950s pre–women's liberation script. She feared her son would lose this fine father figure. But one million dollars was enough money for her to quit her job. What would she do?

She gave the check back. Her son came first. She married the average guy, and they all lived happily ever after.

I sat in the dull glow of the television. In the darkened room, empty paper bag at my feet, I had hoped for a different ending. Why couldn't she keep the million dollars and still marry the guy, so the boy could have a family and all the fun stuff a million dollars could buy?

But what did I know? I didn't like money.

SLEEPING ON IT

1994

His money is twice tainted: 'taint yours and 'taint mine.—*Mark Twain*

I visited my parents in their apartment early one fall evening a few months after Uncle Harry had been admitted to the Sephardic Home. They had just returned from their daily visit to the nursing home, where they ate lunch with Uncle Harry and spent the afternoon with him. They told me the food was excellent. I visited Uncle Harry on Sunday mornings but never stayed for lunch.

Later that night, I was going to attend a seminar on estate planning sponsored by the Brooklyn College accounting society. The speaker was my old tax professor, Harold Solomon. I figured I might learn something, get some continuing education credits, which I needed to maintain my CPA license, and see my old teacher.

I had told my parents I would stop by to see them before the seminar. I sat down on the familiar worn-out green corduroy bedspread that still covered my old bed in the dinette. My mom's favorite color was green. Not only were the living room walls mint green, the foyer's linoleum floor was lime green, and the dinette was stenciled with forest green leaves.

I noticed a well-worn dime on the bedspread. But Franklin Roosevelt's image was not on its face. I had never seen a dime like this before. I checked the date. And then I checked the date again. 1836. I called my dad over.

"Dad, did you see this dime? It's from 1836. Where did it come from?"

"We found some coins in your uncle's apartment, but most were fairly new. It must be 1936. It can't be 1836."

"Dad, this dime is from 1836. See for yourself."

Dad's eyesight was going. He turned on the lamp on my old desk and held the dime within inches of his face, squinting through the eyeglasses I had purchased for him when he was in the hospital.

"You're right. It's 1836. I thought it was 1936."

"Dad, why don't I take that dime somewhere to find out what it's worth?"

He agreed and gave me the coin. I later found out the dime was worth only about ten dollars. But it made me wonder what else was buried in Uncle Harry's apartment.

I sat down at the dining table with my parents. Since his discharge from the hospital, Dad had recovered rapidly. I felt he was strong enough to answer a question I had wanted to ask for a while. I asked it without hesitation. I felt I had a right to know.

"Why didn't you ever tell me about the money?"

Dad's tone became apologetic. "It wasn't our money."

It wasn't *his* money is what he should have said. My mom certainly had both a legal and a moral right to her share of the dough. I looked at Mom. She didn't say a word. The silence made Dad even more animated. Dad told me that ever since Uncle Harry could no longer manage his own funds due to Alzheimer's, he had controlled and managed Uncle Harry's assets.

"We never touched a dime," he said proudly.

The tone of his voice reminded me of Mom's when she used to tell me I didn't like money. I didn't press him. I didn't want to embarrass him. But now I finally understood. Nothing had changed since we were all in the Garden thirty years before. Dad was still too proud to let Uncle Harry pay for dinner, and charity didn't begin at home for Uncle Harry. It never had. My poor mom was stuck between my dad's foolish pride and her skinflint brother. And though I had never known it till then, I had been drowning right next to her. Dad was too proud ever to ask Uncle Harry for financial help for anything.

Now that I knew about the money, I was going to do all I could to make sure history did not repeat itself. I mentioned the seminar I was attending that night on estate planning.

"You know, this is hard for me to bring up, but I want you both to be aware of what is going to happen when Uncle Harry dies."

"He's strong," Dad insisted. "Harry is going to live another five years."

"That may be true. But we want to avoid the situation you must have had when Uncle Joe died. Half the assets in Joe's estate must have been paid in estate taxes."

"No, not half. His assets were held jointly with Harry, so only half the total assets were included in his estate."

"But wasn't that still over a million dollars in estate taxes, Dad?"

"Yes, it was. I prepared the estate return."

"The only significant planning we can do now is planning for after Harry dies since he lacks the capacity to make financial decisions at this point."

"His will and trust leaves everything to your mother."

"Mom, when Uncle Harry dies, the estate liability will be millions of dollars. And then, when you die the estate will be halved again."

"So what can we do?" she asked.

"When Uncle Harry dies, you should consider filing a disclaimer. It's a standard part of postdeath planning when no action was taken during the decedent's life to reduce the estate taxes."

"So how does this disclaimer work?" Dad asked.

"A disclaimer is a legal way for a beneficiary of an estate to say 'thanks, but no thanks' for any part of the assets he or she will inherit. The assets the beneficiary disclaims pass as if the beneficiary predeceased the decedent. In other words, if, when Uncle Harry dies, Mom disclaims a portion of the estate that passes to her, those assets will pass as if she died before Uncle Harry. Those assets will pass to me."

"Do you have to disclaim the whole estate, or can you do just part?"

"Just part is fine. It's up to you both how much."

Dad said he understood. Mom nodded. But neither said what they would do when the time came.

"Let us sleep on it," Dad said. "Now, come with me. Our mattress fell off the bedframe's metal supports, and I can't put it back on."

I walked into their bedroom. The right side of the mattress was on the floor; the metal supports on that side of the frame were bent. I bent them back, got a screwdriver, and tightened the frame's supports. I tilted the right side of the double mattress. It was a cumbersome job for one person, and I ended up turning the mattress all the way on its side.

Through a small hole on the side of the mattress, its contents poured out. Coins pinged onto the wooden floor and rolled all over the bedroom. I doubled over laughing hysterically. My family was certainly different; they stuffed mattresses with coins—not bills.

Before I left for the seminar, with the mattress back in place, I convinced my parents they should let me take the coins home and sort through them to see if any were worth saving. As it turned out, very few coins were worth more than face value.

I asked a question about disclaimers that night at the seminar. On my drive back to Woodmere, I wondered if Mom really understood what I had said about the disclaimer, or did she just think I was greedy and *young yet*?

KEEPING FLORIDA GREEN

1960s–1980s

Farce is tragedy played at a thousand revolutions per minute.—*John Mortimer*

It was rush hour at the Store.

A girl sporting an NYU sweatshirt whispered to her gawking parents while a regular customer named George maneuvered around them. On Friday afternoon, Shabbat raced in with customers clamoring for challah. Uncle Joe called out, "Next."

"No, thank you," said the girl.

Uncle Joe's bushy gray eyebrows went vertical. You could hear the crash as the gauntlet hit the floor. A disgrace. No one ever left the Store empty handed.

Uncle Joe's icy stare compelled the girl's father to explain.

"My daughter goes to school in the Village and brought us to see the place. There are no bakeries like this where we come from."

"If you don't mind my asking, where may that be?" Uncle Joe asked through his smile's missing tooth.

"Scarsdale," the girl's mother said.

"Oh, I see. Bakeries way up there in Westchester like spectators, but here we prefer customers. You want a show, go uptown to the Palace!"

Uncle Joe didn't bother to wait for a reply. "So maybe I can interest you in our delicious marble cake? It's fresh and moist; see how wide the veins of chocolate are?" He tilted up one edge of a large rectangular metal tray on top of the raised counter.

The father smiled, "All right, we'll take some."

I can still hear my mom say, "No one moved the merchandise like Uncle Joe."

BY SPECIAL DELIVERY

Hill and Dale Inn
Route 8, Otis, Mass
26 September 1967

Dear Harry,

I hope you are in good health. This time I want a favor and also to register a complaint. I would like you to send me the original twin Babka (two in a box). This other thing (Acme) just creates problems for me it crumbles and breaks apart even when frozen. I have a feeling that the other Babka costs more but send it anyway. I know what I have to work with. End of complaint.

Now for the favor. Please go to the Schwartzes' & ask Mrs. Schwartz or Albert to send at least 5 yards of the enclosed gold braid. This is their product.

I am sending cheque for $13.00. This should be enough. If not we will work it out later. Say Shalom to the Schwartzes & Joe and your friend from the Candy Factory.

"Goot Yantiff and Sie Geszundt."

Hassan

On the back of the letter, Uncle Harry had written the following in pencil:

	Sept 27	
Babkas	7.20	Bal Due
Postage &	13.00	
Fav	2.00	9.20
		3.80

ꕥꕥꕥ

Through the crowd, my mom noticed a young man standing outside the Store in ragged clothes, wearing an expression, as Murray Kempton once wrote, of someone who had "played a game and lost." My uncles knew his parents. They lived in the neighborhood, but their son lived on the streets. Why? No one was sure. He just did.

A nonpaying regular, he was too self-conscious to come in to ask for food, but his inaction was the fulcrum on which a Ninth Street tradition swung into action. Mom sliced a small pumpernickel, bagged it and a few donuts, went outside, and handed him the white bag. He thanked her and turned away.

ꕥꕥꕥ

30 August 1982

Dear Helen, Joe and others at the 9th Street Bakery!

Greeting from Europe! On the reverse side you see the daily farmer's market on the town square of Bonn, the capital of Western Germany. The weather has been wonderful here all summer long, which is exceptional. But the economy is not so wonderful, there is lots of unemployment.

Best wishes,
Chris Klingenberg

P.S. I'm the fellow who comes with the bike
and asks for day-old rye at reduced prices.

Another Friday regular trudged down Ninth Street. An ancient babushka with a round shape, her head beneath a kerchief, wearing a worn coat the color of an eggplant, she tottered on orthopedic shoes. After peering carefully through the picture window, she stepped inside the Store and waited her turn. My mom caught sight of her out of the corner of her eye. As usual, Mom thought the old woman looked like she had just stepped out of the shtetl. But the old woman was brave enough to ask for something for nothing, as she had each week for years. No standing outside in the cold waiting to be noticed for her.

She listed into the limelight before Uncle Joe. As always, he sympathized with her and bottled his caustic wit. When he gave her a challah, he must have thought of all the food packages Lena had paid to have sent back to her relatives in Zetel, Russia—the old country.

Max and Lena Wolkirmerski had left Zetel in 1913 with their boys, Joseph and Harry. Eventually, they boarded a vessel named *Tsar* bound for New York, where, by the time they left Ellis Island, they had received an Americanized name—Wolk.

Max and Lena started with an Orchard Street pushcart, graduating to grocery-store ownership under the shadow of the Williamsburg Bridge, and somehow scraping together the purchase price for a commissioned bakery and lease at the corner of Stanton and Allen Streets, until Moses (Robert Moses, not the guy with the beard) decided Allen Street needed widening and forced a move uptown. In 1926, my grandmother chose a location on Ninth Street. Lena not only helped run the bakery but also did the cooking, laundry, and cleaning at home. After a baby girl was born, they lived the American dream and moved from cramped Clinton Street to what were then the suburbs, the East New York section of Brooklyn.

Just as the old woman left with her usual bounty, a new player entered the Store and gave the old lady a big hello. She exited hastily without acknowledging him. Uncle Joe took notice, surprised she could move that quickly.

"Excuse me, mister, you know the old lady who just left?"

"She's my landlord; she owns half the buildings on Tenth Street."

"Well, what do you know," said Uncle Joe.

ꟹꟹꟹ

A Charlie Brown Thank-You Card

5 Cornelia Street 2C
New York, N.Y. 10014
22 July 1969

Dear Harry and Joe,

Now I am quite contented, yet I do miss the nice feeling of greeting you or having a chat, and getting good bread on those occasions I just had to ignore my weight.

So, if belatedly, I do most genuinely thank you for your valued assistance and boxes and string when I was moving, and know that all your kindnesses and thoughtfulness are truly appreciated.

See you soon—

Warmly,
Polly

ꟹꟹꟹ

Uncle Harry returned from a delivery. "Hello, Joe. Hello, Helen."

In response, Uncle Joe rolled his brown eyes toward the front door, where a familiar New York City health inspector was about to enter. "Harry, look."

Uncle Harry scooped up the sleeping Suzy from her box on top of the radiator and stuck her under his coat while Uncle Joe kicked her litter box under the chair in the back room. A stream of sawdust, doubling as cat litter, now stretched across the floor. With his left hand, Uncle Harry opened the front door for the inspector, holding Suzy behind his back under his coat with his right hand.

Mom greeted the inspector with upraised arms. "Hello! What can we do for you? I have some wonderful seven-layer cake with your name on it right here."

The inspector turned to my mom. Uncle Harry nodded hello to him while he backed out the front door. Uncle Joe quickly slid some boxes in front of the litter box beneath the chair. Before the inspector could answer Mom, Uncle Harry was back in the Store. Suzy meowed in the trunk of Uncle Harry's car double-parked in front of the Store.

You did not have to see the litter box to know a cat lived in the Store. There was no way you couldn't smell it. But years later, when I picked up the gnawed, brittle, paper-trail remains of a twentieth-century New York City immigrant family that never threw anything away, I did not find a single citation for health violations, only lots of checks to the New York City Parking Violations Bureau. No one ever left the Store empty-handed—not even health inspectors.

With the inspector gone, the old lady from Tenth Street became the subject of conversation.

"Harry, you know that old lady who comes in before Shabbes every week? Mama used to give her bread for nothing?"

"Yeah."

"She owns half of Tenth Street," said Uncle Joe.

"You don't say."

"A new customer came in and saw her. She ran like mad. He told me she's his landlady."

"So from now on we charge her. A *gonif.* A thief. She even had Mama fooled."

"She won't be back, Harry. Our prices just got too high for her."

❧❧❧

15 August 1985
LE CINQUE TERRE-RIOMAGGIORE

Dear Harry,

Greetings from the Land of Pasta and Vino. We have been to Austria, France, Germany, Spain, Switzerland and still we haven't tasted bread as good as yours! We look forward to seeing you soon and again tasting your Russian black bread.

Love
Jay + Judy Mancini

❧❧❧

In the early 1970s, congressional elections brought a flurry of activity at the neighboring Democratic Club. The members ordered both cherry and cheese Danish from Uncle Harry for the occasion. Their special visitor always stopped by the Store to shake hands with my uncles. Uncle Joe referred to her as the lady with the big mouth and a hat to match. My mother doesn't recall if Bella Abzug, the New York congresswoman, favored bread or cake, but Mom assures me that even Bella didn't walk away from the Store empty-handed.

❧❧❧

"Joe, you know who I just saw?"

"Who?"

"Arnold, outside his health-food store at the corner. He says business is good, and he and his *mishpocheh,* family, are thinking of investing in noncarbonated drinks like tea."

"Hot tea?"

"No, iced tea in a bottle."

"Who is going to buy cold tea?"

"I don't know. I think he should stick to what he knows."

When I was six, in the pantry of the house on Hegeman Avenue, Uncle Joe taught me how to drink hot tea from a glass by placing a cube of sugar in my cheek. The rear pantry room behind the kitchen was always sunny. It overlooked a cherry tree. The two of us sat there at a big, shiny metal table in front of two tall glasses, wisps of steam, and a sugar bowl.

"This is how you drink it, Morton. You put the cube of sugar in the side of your mouth like this, pick up the glass, and sip slowly."

"It's hot, Uncle Joe. Why can't we use a cup? It's easier to hold."

"No, Morton, a glass. It tastes better from a glass."

For Uncle Joe, the tea was never hot enough.

Years later, just about when Uncle Harry put a lock on the front door of the Store for the last time and "gave away the key" to one of his suppliers, Arnold and his relatives sold their natural iced-tea concept and the business they started with that idea for "a very pretty penny," as Uncle Harry used to say. They called it *Snapple.*

❧❧❧

305 E. 11th St.
New York, N.Y.
2 February 1986

Dear Harry:

I heard that you hurt your leg and I was very sorry to hear that. Every time I go past the store and see that it is still closed, I get a pain in my heart.

I hope and pray that you are not closing the store for good. I can't bear to see the best stores on the Lower East Side closing—and all being replaced with such garbage—art galleries and boutiques, all selling junk, when what we need is bread, milk, drugstores—*real* stores, selling to real people—not pornographic art, selling to uptown people slumming in our neighborhood.

Please let me know how you are, and I hope you are planning to reopen the store. If you need help, please let me hear from you.

Love,
Frances

❧❧❧

George stepped into the breach and ordered a large challah from Mom.

"How are you, George? Where are you going on your next vacation?"

George always mailed postcards to the Store when he was traveling. He was not the only one. A collection had already started to gather dust in the back room: hundreds of postcards and letters from customers, both near and far, as they traveled all over the world in search of the ultimate babka. Uncle Harry asked the customers to mail postcards to the Store. I guess Uncle Harry lived vicariously through the customers. Through the years, George's scribblings from St. Moritz, the Acropolis, and Semmering all found their way to Ninth Street, where the longest trip my uncles ever took was back to Brooklyn to pick up stuff at Pechter's, RK, or Danilow.

❦❦❦

Postmarked July 21, 1966 POSTCARD
FROM: TO:
The New South Seas Hotel Joseph & Harry Wolk
On the Ocean at Eighteenth Street 350 East 9th Street
Miami Beach, Florida New York, New York

Hello!

How is everything? We still don't know how we'll come back: plane or bus.

Write.

Regards—
Helen, Phil & Morton

❦❦❦

Just because Uncle Harry and Uncle Joe didn't go on vacation, didn't mean Mom, Dad, and I didn't. Every summer, from the age of six until I was twelve, we journeyed to Miami Beach, where low hotel prices offset the high temperatures and humidity.

In the 1960s, the name "South Beach" did not exist. The beautiful people flocked elsewhere. Each summer we stayed at the same oceanfront hotel at Eighteenth Street called the South Seas. The place was effectively a Jewish old-age home. Run-down turquoise-colored Art Deco hotels like the South Seas, whose heyday had taken place decades earlier, swarmed with retirees. The South Seas' claim to fame was that the Alaskan delegation to the 1968 Republican convention had stayed there. That probably tells you all you need to know about the South Seas Hotel.

There were rarely any kids for me to play with, so I spent hours by myself in the ocean throwing myself against the waves over and over, pretending I was Dick Butkus, the Chicago Bears middle linebacker, tackling opponents. Other times I walked up and down the beach looking for seashells with another hotel guest named Bea, who was old enough to be my grandmother. At night my parents and I would take a walk, or get on a bus, and visit some of their friends. I recall visiting a lonely old widower whose wife, I was told, had died just a few years after he sold his delivery route, retired, and moved to Miami Beach. He lived at the Thunderbird Hotel and spent so much time outside in the sun that his deeply tanned skin looked like leather. His white hair reminded me of snow; or perhaps I felt that way because the story of his winter driving accomplishments had preceded him. That was the one and only time I met Mr. Martin Cohn.

We usually flew to Miami, but in July 1966, when my mother sent her *regards* to her brothers, there was an airline strike and we took the train. I don't recall if it was in the train station in Miami that year or another summer at the Miami International Airport, but in one of them, the sign greeting you on arrival read:

KEEP FLORIDA GREEN; BRING MONEY

My parents laughed out loud at the sign. Dad tapped me on the shoulder to make sure I saw it. I got the joke but did not think it was that funny. I was more interested in convincing my parents to give me some change for a vending machine that made plastic alligators while you waited. They kept Florida green and bought me one.

GENERATIONS

1995

Everything revolves around bread and death.—*Yiddish saying*

The day Uncle Harry died, I stood on a levee on the western bank of the Mississippi River consoling my four-year-old son, Ari. We had no idea Uncle Harry was dead; Ari had just slipped and skinned his knee. We had taken a brief rest stop at a historic plantation on our way to the airport for our return flight to New York. Our infant daughter, Aleeza, slept in Nurit's lap. Nurit, Ari, and I had flown from New York to New Orleans two days before, rented a car, and drove north to bring our two-week-old adopted daughter home. After driving for miles over the bayou, an endless, eerie display of isolated trees growing out of muddy water, the stop gave me time to reflect on my first moments of fatherhood.

Four years earlier, after years of struggle with infertility, Nurit and I had adopted Ari in an even more rural part of the Deep South. "Rural" is an understatement. We drove for miles on that trip and never saw a sign, except for the occasional "Fresh Dung—ten miles." At the lawyer's office, where Nurit and I signed papers, an Ole Miss football helmet sat on the desk. The helmet made me realize how far we were from New York. We expected to go to the adoption agency to get Ari after the paperwork was in order. But as soon as the final signatures were inscribed, a crying baby boy was placed in my arms.

Miss Essie, sporting a southern accent heavier than the Mississippi Delta humidity, had cared for Ari from the time he left the hospital two weeks before until that moment.

"Y'all knows, Mr. Zachter, when a baby cries, it's their way of saying they need one of three things—feedin', holdin', or changin'. Now, I just fed that baby of yours, and y'all holding him pretty good, so I'd say you're at number three."

"Oh."

"Mr. Zachter, this being your first and all, let me ask you, have you ever changed a baby before?"

"No."

"Well, why don't y'all just step over to this here table."

She held up a disposable diaper in one hand and a towel in the other.

"Y'all always want to put something underneath the baby before you put him down."

She handed me the towel and the diaper.

"Uh-ah, that's right, towel down first, baby on top, and then y'all remove the old diaper. Y'all pull down those tabs on either side and then lower the front of the diaper slowly."

Ari decided it was as good a time as any to introduce himself. He peed up like a fountain. My shirt soaked, I started laughing.

"Mr. Zachter, y'all wants to take that front part down real slow next time if y'all wants to stay dry."

"That's okay. I've been waiting for this for a long time."

Ari was an active participant in Aleeza's adoption. Every time we stopped our drive and started again, he insisted on strapping Aleeza into her baby seat. Our flight to New York was the next morning, so we took the scenic route back to New Orleans through the bayou. That night we stayed at a hotel in New Orleans overlooking the Superdome. Ari said someday he would play football there, and Aleeza would be a cheerleader.

We didn't get much sleep. There was a convention going on in the hotel, and it was also our first night with Aleeza. By 10:00 p.m. I had given Aleeza her nickname: "Squawk Box."

Nurit figured my mother-in-law might know how to put howling baby girls to sleep. Rumor had it that Nurit had not been the quiet type either. When we called Ellen, we learned that Uncle Harry had died that morning, just about the time we picked up Aleeza. The funeral was the next day, and our flight would not get us back in time. I felt guilty knowing I had put Uncle Harry into that nursing home, which probably shortened his life, and now I wouldn't even be at his funeral. My only consolation was that we had kept him at home for as long as we did. He lived in the nursing home for only four months. I'm sure the four months seemed a lot longer to him than they did to me, visiting him only once a week.

The children and Nurit finally fell asleep from exhaustion. I wandered through the lobby and tried to remember the Uncle Harry of my youth: the uncle I loved, the uncle who always had a joke to tell, the one who asked me to plant bushes with him. But the selfish hoarder of recent memory was what came to my mind. As you sow, so shall you reap. Uncle Harry knew that Nurit and I had to take a second mortgage on our house to adopt our son, and he never offered to help despite his millions. He even had the honor of holding Ari down during the bris (circumcision). And when it came to adopting our daughter, when we already had two mortgages on our home and I was struggling financially, Nurit's uncle and aunt lent us the money while my Uncle Harry sat on millions.

Uncle Harry's selfishness continued even in death. No estate planning had been done; half of everything he owned would be paid to the federal government in estate taxes. I was a CPA with a specialty in taxation. I had taught tax at NYU. And no one in my own family had come clean and gotten my advice. The estate tax would be in the millions. There was no avoiding it; even if my mother filed the disclaimer, which was far from certain, since I was still receiving signals that I was *young yet,* that would save only future estate taxes. I wondered how much the deposit was on an F-16 fighter plane. The estate would probably receive a thank-you note from the Pentagon.

And then it hit me. *Shouldn't estate taxes be done away with? They are immoral. Who invented this stuff?* I had been a Democrat all my life, but in the course of one phone call, I started to feel differently. I must have set some kind of speed record. Quicker than you can say maximum federal-estate-tax rate, I started to think like a conservative Republican. But after a few moments, I pulled myself together. I would remain a registered Democrat.

The next morning we flew back to New York. At the same time, in Queens, only a few miles from the final resting place of New York's most famous hoarders, the Collyer brothers, Uncle Harry was buried next to Uncle Joe.

Aleeza slept the whole plane ride home. She didn't make a sound.

LEARNING TO FLY LOW
1960s

In the long run you hit only what you aim at. Therefore, though you should fail immediately, you had better aim at something high.—*Henry David Thoreau*

Memories of fall always bring me back to my uncles' apartment for the two days of Rosh Hashanah. The Store closed for the High Holidays, and Uncle Joe left for shul early in the morning; going to shul was his version of a vacation. By the time I woke up, Uncle Harry and my parents were in the dinette finishing breakfast. My arrival always brought out the same reaction from Uncle Harry.

He took one last sip of his coffee, which he drank with heavy cream and three packets of saccharine, about which my mom always said, "What's the matter, you're not sweet enough?" When I made my appearance, Uncle Harry jumped up from the table, removed the floral-patterned tie tucked inside his shirt, grabbed the small black frying pan from the stove top, and twirled it. "Morton, you want me to make you eggs like I used to make in the goulash joint?"

Then he would laugh, a contagious laugh that spread first to my parents and then outside through the open window to linger in what, in my mind's eye, was always a hot and humid Indian-summer morning on which the breeze off First Avenue blew back into the apartment as if from an oven. Uncle Harry's brown eyes came alive. His head nodded up and down. His hair was slicked back, and he reeked of aftershave. He wanted a yes from me. And I never disappointed him. On those Rosh Hashanah mornings, I ate ketchup and scrambled eggs for breakfast—in that order.

Uncle Harry always spouted funny stories, so there was no need to ask him about the goulash joint. It was just another of Uncle Harry's jokes that only the adults understood. But even as a child, I sensed something in the pleasure he got in asking that suggested Uncle Harry had done a lot more than scramble eggs in that goulash joint.

After breakfast, in her white blouse and plain, dark, long skirt, Mom cleared the table, washed the dishes, scrubbed clean the frying pan Uncle Harry had used, and left them all to dry on the plastic dish rack beside the sink.

Then she took off her apron and said, "Let's go. It's late."

The four of us walked downtown on First Avenue. At Houston Street, we stopped for the red light. I read the street sign, pronouncing it like the city in Texas.

Mom corrected me. "Morton, what's with you? Are you a cowboy? House-ton Street, Morton, House-ton."

I couldn't figure out why the city in Texas was spelled the same way as the street in lower Manhattan but was pronounced differently. As a child, I decided that certain words were pronounced correctly only in New York. Today, I know that Houston Street was originally called Houstoun Street, named for William Houstoun, a delegate to the Continental Congress in the eighteenth century; but sometime in the intervening centuries, the name metamorphosed into its current spelling despite retaining the original pronunciation.

We crossed the pedestrian island on Houston, sidestepping

shards of broken bottles the color of faded negatives, past men with worn leather faces sleeping on cardboard beds. The top layer of one man's newspaper blanket floated off in the aroused humidity masquerading as a morning breeze. Our nostrils filled with the smell of sweat and urine as we reached Allen Street. As always, we stopped at the corner of Allen and Stanton Street to pay homage to the past.

"You see that spot, Morton?" Mom asked. "The Store was there until 1926."

She pointed with her left hand to the empty pavement on the northbound side of Allen Street. Despite the heat, she wore a long-sleeved blouse. On the inside of her wrist, her watch band was stuffed with white tissues folded in neat rectangles.

"Right there, you see?"

I could tell this was important to her. I made sure to stare at the air where she pointed and to give the impression that I saw my grandparents serving customers almost a half-century before.

Uncle Harry said, "Someone came in and offered to buy out our lease. Your grandmother said yes. It sounded like a good offer at the time. What did we know? A few months later, the city bought out all the leases fronting on Allen Street for a much higher price, to widen the street and put in that center island as a pedestrian park."

"And that was 1926, when your grandmother looked uptown and picked Ninth Street as the best location for the Store," Mom added.

With this, Uncle Harry turned and led our march a block east to the second and final required tour stop on our way to shul. We stood in front of 85 Stanton Street, where United States senator Jacob Javits was born in 1904. A rectangular brass sign on the tenement indicated his birthplace.

"Another child of Russian Jewish immigrants who started with nothing on the Lower East Side and made it big," Uncle Harry said.

Mom added, "When President Kennedy, may he rest in peace, came to New York, Javits took him here to show him where he came from."

"Javits wasn't born with a silver spoon in his mouth. He didn't know from Cape Cod. Cod liver oil, this he knew," said Uncle Harry. "They didn't have hot water in these buildings or an elevator."

"Cold-water flats," said Dad. In his blue suit and well-shined shoes, he nodded.

I figured I was lucky. We had hot water in our tenement building in Brooklyn. At that time, cold showers seemed a lot worse to me that sleeping in the dinette with my head next to the refrigerator.

Over thirty years later, I returned to look for the Jacob Javits sign, but it wasn't there. It now exists only in my memory. Perhaps it had been stolen and never replaced; or perhaps the sign was never there—I had imagined it—and my family just knew the building by sight.

No one but my mom was in a hurry to get to shul and have Uncle Joe glare at us for arriving late. So we traipsed down a deserted Orchard Street, window-shopping. The Jewish immigrants who had lived here, and overcrowded this street before World War I in search of bargains from the pushcart vendors, had long ago graduated to Brooklyn, the Bronx, or beyond. But the retail stores were still Jewish owned, and this being a holiday, they were all closed.

At that time, the latest wave of New York immigrants lived in that neighborhood. Some Puerto Rican kids were playing stickball on the empty street, stopping only when the occasional car passed. I watched one kid swing a cream-colored broom pole with black construction tape twirled around it. It looked like he was swinging a barber pole. But he hit the pink rubber Spaldeen, and his team screamed, "Run!" I wished I could play. But dressed in my short-sleeved white shirt, clip-on tie, and dark wool blazer, I had a higher calling.

The white brick facade of the First Roumanian–American Congregation came into view when we turned left onto Rivington Street. Founded in 1885, this had been my grandparents' shul when they lived on the second floor of their cold-water flat at 30 Clinton Street.

We enjoyed the shade of the windowless shul lobby, relieved to be out of the heat. Directly ahead of us was a door leading to the tiny downstairs sanctuary used on most Shabbats. But Rosh Hashanah was a major holiday, and the sizable crowd required the large main sanctuary upstairs. Uncle Harry told me that even the goyim go to shul on Rosh Hashanah. What he meant was, even Jews who never went to shul showed up on the High Holidays. He fit that category, as did most of the congregation, which was composed of the few elderly Jews still remaining on the Lower East Side. They, in turn, dragged their children and grandchildren, reeled in for the holidays from suburbia, to shul with them.

We walked up to the second floor, used exclusively by men. The First Roumanian–American Congregation was Orthodox, and the women were not allowed to sit with the men. My mom continued up to the third floor, where the women sat in a U-shaped balcony. As we watched her trudge up the third flight of stairs, Uncle Harry referred to the separate seating as a *mechaieh* (a relief) for the married men. It was also beneficial for him since my mom couldn't sit next to him and harangue him about closing the Store on Saturdays.

Uncle Harry creaked open the door to the sanctuary. Heads turned to see who was so late. We scampered to the same row we always sat in, on the right side of the sanctuary toward the back. It was after ten. In his brown suit, Uncle Joe gave us The Look. He didn't say a word. He didn't have to. The Look was enough.

Uncle Harry sent me into our row first to act as a buffer between him and Uncle Joe. I didn't mind. Uncle Joe's Hebrew skills were better than Dad's or Uncle Harry's. He opened the prayer book to the correct page for me and pointed to the line the cantor was singing. The cantor sang and chanted at a rapid pace, the congregation mumbling right behind him. My Hebrew skills were still in the formative stage, so, within seconds, I was lost.

I spent the next hour staring at the inside of the shul, sweating and wishing I could take off my blazer. All the windows in the sanctuary were open, letting in the heat. Colorful stained-glass windows lined both sides of the *bemah* (the raised area in the front of the sanctuary) with its ornate mahogany woodwork. On the woven curtain covering the doors to the Aron Kodesh, where the Torahs were kept, embroidered lions stood upright on their hind legs holding the Ten Commandments. The congregants also held my attention. I decided those congregants who *shuckled* and mumbled a lot must be paragons of religious observance. Some even put their *talaysim* (prayer shawls) over their heads to act as blinders so they could focus on their prayers.

Not Uncle Harry. He leaned over and constantly pointed out to me who the big shots were. If he liked them, he called them *machers*.

"There's Feldman; he's a lawyer. Behind him is Schwartz; his son's a doctor."

If they were rich, they were big in Uncle Harry's eyes. Those congregants Uncle Harry did not care for he referred to as *k'nockers.* He knew many of the congregants from the Store. If they chiseled too much on their babka purchases, they were *k'nockers.*

When Uncle Harry was not whispering in my ear, I loved to listen to the choir. Six men in white robes called kittels, baritones and basses, sang ancient melodies with religious fervor, sweat pouring from them. Their incantations were timeless; I imagined Jews singing these same melodies hundreds of years before. They stood in two rows of three on either side of the cantor. When they paused, I watched their robes heave up and down as they tried to catch their breath. They were the same six men every year. My favorite was the last one on the right of the cantor. He was tall and thin, with a huge milk white forehead that contrasted with his dark hair and brown eyes. As he sang, he smiled and made eye contact with the other choir members. He sang with his entire body, and I saw the joy he felt in being where he wanted to be, doing exactly what he wanted to do.

Every once in a while, we stood up, then sat down, then stood up again and sat down. This up-and-down action served an important purpose for me. It kept me from dozing off in the heat. When I stood up, I was afraid one of the *machers* in a nearby pew could tell that I had no clue where the cantor was in the service since my prayer book was open to the wrong page. The service was conducted entirely in Hebrew, but the prayer books had Hebrew on the right side of the page and the English translation on the left. Dad gave me comfort.

"Read the English," he said.

After a while, Dad grew bored and whispered to me, "Why don't we take a break."

I was up in a flash. Out in the hallway on the second floor, I killed time reading the names of the congregants who had made donations to the shul decades before. Their names were engraved on huge marble plaques attached to the walls. At the turn of the twentieth century, you got your name on the wall forever for only ten dollars. This impressed me. Dad explained that ten dollars in 1910 would be worth hundreds or perhaps thousands of dollars today. The concept of the time value of money made a lot more sense to me than why Houston Street was pronounced house-ton.

I put my index finger in the carved marble indentations and pretended to write the amounts. The marble was cool to the touch. I reached up to touch the bottom two names: Binder Sarah $10 and Cohan Zlata $10. High above my head, at the top of the slab, Fred Siegler $250 remained unreachable. I wondered why Binder Sarah and Cohan Zlata were listed last name first, but Fred Siegler was listed first name first. I figured Fred must have been a *macher.*

Dad and I stood outside the shul, the fierce midday sun blazing down on a Rivington Street that was still damp from a sanitation sweeper wash and whose pavement glittered. Across the street, a store that sold Jewish prayer books, *talaysim,* and mezuzahs was closed, silent in the shadows. I walked to the corner and watched the stickball players' faces glisten with sweat, while my dad, a habitual smoker, resisted the urge forbidden on this holy day. After a while, we strolled back inside for the rabbi's sermon.

Every year on the first day of Rosh Hashanah, Rabbi Mordechai Meyer gave his speech in Yiddish, and I had no clue what he was saying. But I knew it was meaningful from the way everyone nodded his head in agreement. I watched Uncle Joe's face as he soaked up every word. He greatly admired a man as well versed in Torah as Rabbi Meyer. Uncle Joe would smile proudly when he had a chance to shake the rabbi's hand and wish him a happy holiday. But this was the second day of Rosh Hashanah, and the rabbi always spoke in English on the second day.

Rabbi Meyer stood up from his seat on the right of the *bemah* and leaned on the brown wooden railing in front of him. In his white kittel, his short, squat body and fleshy face reminded me of a resplendent bulldog. But I felt an affinity for Rabbi Meyer because we shared the same Hebrew first name.

The congregation continued to talk after the rabbi stood up. From his seat on the *bemah*, the red-headed president of the shul slapped his hand down hard on the top of the wide wooden handrail. The thud silenced the congregation. Everyone was still. The rabbi began to speak just as an ambulance's siren wailed by outside. Despite the congregation's annoyance, he waited patiently for it to pass, as if he understood that the wailing ambulance represented a far greater suffering. And then, from the silence, his voice boomed through the shul with an urgency charged by conviction. The man did not need a microphone. He pointed to the window open high above him at the front of the shul.

"A little bird flew into the shul through an open window," he said.

"After flying around the ceiling, it wanted to get out and flew in the direction of the window. But the little bird flew too high. It hit the wall between the top of the window and the ceiling. Over and over again, the little bird hit the wall above the window. It became frantic. Banging its beak wildly into the wall, it bloodied itself and fell to the floor. Dead.

"What a shame. Why should this have been? There was no need. If the little bird had only flown a little lower, it would have made it out the open window. It would have been free. It would have soared.

"Do not be like that little bird. If you want to be happy, do not forget what is within your reach. Set your sights lower; be reasonable in your aims—your goals—so you can fly. Do not reach so high; fly a little lower in all things. Do not hold in awe that which you cannot acquire. Fly low so you can soar contented."

The congregation loudly replied in unison, "Amen."

The rabbi's speech was the center of our conversation as my family and I walked back to my uncles' apartment. Uncle Joe summed it up for my benefit. "Did you hear, Morton, what the rabbi said? Don't have big eyes."

COUNSELOR ZACHTER

1995

A lawyer is a learned gentleman who rescues your estate from your enemies and keeps it for himself.—*Lord Henry Brougham*

The building in downtown Brooklyn was formal, the drapes red velvet, the New York State flag prominent, the soon-to-be lawyers already jaded, and the ceremonial calling of several hundred names endless. On the morning of January 18, 1995, before a black-robed judge from the Appellate Division of the Supreme Court of the State of New York and my smiling parents, I was affirmed as a member of the New York Bar, having been found to be "a person of good moral character" after four long years of attending law school at night and passing the bar exam.

Afterward, my parents and I strolled down Joralemon Street to the Kings County Surrogate's Court, located at the western end of the New York State Supreme Court building. Despite its name, which is intended to confuse out-of-state lawyers, the New York State Supreme Court is the lowest court in the New York State system. The full spectrum of humanity that makes Brooklyn great eventually finds its way there when it feels litigious.

Fast-moving revolving doors swooshed us into a lobby crowded with defendants, plaintiffs, and lawyers. The din bounced off the stone walls and concrete floor and echoed from the high ceiling. I spied two empty seats on a wooden bench against the wall of a hallway off the lobby, sailed my parents through the legal sea of flotsam and jetsam, and sat them down.

I said, "I'll be back in a little while. You wait here. Afterward we'll go for lunch."

My dad looked up at me. "Do you want me to go with you? After all, I am still a lawyer."

"No thanks, Dad, I want to handle this one on my own."

I walked down the hall and entered the office of the small-estate section of the Kings County Surrogate's Court. I closed the door behind me. The line leading to the gray metal counter manned by a young clerk was short; only two people were ahead of me. I waited patiently. Another few minutes would make no difference. I had been waiting for this for a long time, even before I knew there was anything to wait for. Until only a few months ago, I had still dreamed of winning the lottery. But you have to be in it to win it, and I was never in it. Once upon a time, as a CPA at Deloitte, I had a client who won the New York State lottery, but that's her story, not mine.

Now it was my turn. "Hi," I said as I nodded and handed the clerk one original and one duplicate set of the required papers for the estate of Harry Wolk. The clerk took the papers and read them in front of me. I watched his eyes move down the pages, making sure all the required information had been supplied: original death certificate, a signed will with affidavit of attesting witnesses, properly executed trust document, and the court's preprinted informational forms fully completed. Then he got to the disclaimer.

He stopped. His eyes narrowed. One end of his mouth curled up.

"What is dis for exactly? We don't often see disclaimers here."

"Virtually all the decedent's assets pass by operation of law to his sole surviving sibling, Helen Zachter, in accordance with the trust agreement. Harry Wolk had no spouse, was never married, and had no children. All his assets would pass to Helen Zachter even if he had died intestate and there were no trust agreement. The only assets passing under his will are some sundry bank accounts with a total value of less than ten thousand dollars. This is why the filing is here in the small-estate department, which is designed for estates below ten thousand dollars, if I understand correctly."

"I know all dat. What about dis disclaimer for dees tree brokerage accounts?"

"For estate-planning purposes, Helen Zachter is disclaiming her interest in some of the assets passing to her under the trust agreement. The three brokerage accounts represent those interests and are therefore passing as if she had predeceased her brother. Since this is the only required court filing, I am including the disclaimer as part of this filing. Is the disclaimer improperly drafted or documented?"

"No, it's fine. How are you related to Helen Zachter?"

"I'm her only lineal descendant."

"Den da assets in da tree brokerage accounts being disclaimed will legally pass to you?"

"Yes, that's my understanding."

"But you are also da attorney of record for da estate."

"Yes."

"Wait here."

Before I could say anything, the clerk sprinted to the back of the office and knocked on the door of the senior clerk of the court. The door opened, and a heavyset man of about fifty had his turn to thumb through the papers I had made sure were accurate and complete. The young clerk whispered to him in a voice I couldn't hear. I felt the sweat drip from my underarms beneath my legal armor—T-shirt, long-sleeved white shirt, suit, tie, dark wool overcoat, and the long gray scarf my mom had knitted for me. The senior clerk finally nodded to the young clerk, got up, made eye contact with me, and approached the counter, never taking his eyes off mine. He did not say hello.

All he asked was, "How much is in the three accounts being disclaimed?"

I hesitated. I thought that by answering honestly, I might prolong this matter for months, maybe years. Who knew what hoops they would ask me to jump through in my very own Brooklyn *Bleak House*. But I was now an officer of the court and a person of "good moral character."

All I said was, "Five million."

He smiled broadly and turned to the young clerk.

"Please process the disclaimer for Counselor Zachter." He turned back to me. "I haven't seen you around the court before. Are you new to Kings County?"

"No, I am new to the bar."

"How long have you been practicing, Counselor?"

I looked at my watch.

"About an hour, sir."

"Welcome to the New York State Bar, Counselor Zachter."

"Thank you."

The senior clerk turned and floated back to his office, still smiling. The young clerk stamped my papers, kept a copy for the court, and handed me my set. For the moment, before the estate-tax bill halved it, with my first act as a lawyer, I left Surrogate's Court five million dollars richer. And that was before I collected my legal fee for services rendered.

Back in the din, Dad asked, "How did things go?"

"No problems," I said.

With this, my mom beamed, secure in the knowledge that I would never have to worry about money again and would make sure her grandchildren wouldn't either. Apparently, I was no longer *young*. Though she had disclaimed the majority of the estate, she kept a little something for herself as back pay for a lifetime of moving the *merchandise*. Sadly, it would make little difference in her future life. Living well, or even better, was beyond my parents' expectations and abilities.

We lunched in a coffee shop on Court Street—tuna on whole wheat, with lettuce and tomato, sour pickles, and regular coffee in thick white mugs. We spoke about how beautiful their baby granddaughter was, and how helpful Ari had been last week when we picked her up in Louisiana. The first check I would issue with my newfound wealth was to repay the loan Nurit's uncle and aunt had given us to adopt Aleeza.

I said goodbye to my parents after lunch. They took the subway to Kings Highway, back to their apartment. At Atlantic Avenue, I caught the LIRR bound for my home in Woodmere, where my wife, young son, and infant daughter waited. In the midst of winter, I was convinced spring had arrived.

In time, I came to know better. Money brings comfort, but happiness must grow from within. Like the ailanthus, it requires only water and sunlight.

CHARITY

1960s–1970s

Few sinners are saved after the first twenty minutes of a sermon.—*Mark Twain*

A week after Rosh Hashanah services at the First Roumanian-American Congregation, my parents and I returned to Manhattan for Yom Kippur, the Day of Atonement, a day of fasting and self-reflection. On *erev* Yom Kippur, we ate a lot of chicken for dinner because we couldn't eat until sunset the next day. We also had to eat fast. Unlike Passover, when we didn't go to shul, my uncles had to close the Store before sundown so they could eat dinner at the last possible moment. We always arrived for shul on *erev* Yom Kippur just in time for Kol Nidre, a solemn prayer in which the cantor voids vows to God on behalf of the congregation. We ran down Allen Street like mad not to miss it.

That night was always when the annual Kol Nidre fund-raising appeal was made at the shul. First, Rabbi Meyer gave a speech in Yiddish that Uncle Joe told me had as its major theme the importance of giving *tzedaka* to avoid an unfavorable decree when God closed the books on the next year, determining who should live and who should die. "Charity," Uncle Joe would ask, "you know what that is Morton?"

Atonement was serious stuff. I was twelve. My biggest sin of the past year was when I faked being sick so I could stay home and listen to the Yanks' opening game of the season on the radio. If

I recall correctly, they even won on a Roy White homer. If a bird could get killed for flying too high, who knew what could happen to me for deceiving Mom.

After the rabbi's lengthy speech, members of the congregation stood up or yelled out from their seats how much they were donating to the shul for the Yom Kippur appeal. This was my first taste of adult peer pressure. I thought the whole thing was nuts. For the shy members of the congregation, the shul president walked around and had them whisper to him how much they were donating. He then yelled out in a loud voice so everyone could hear that person's name and the amount. Some wished to remain anonymous. They would not tell him their name.

For these he yelled out, "From a nice man in the back, five dollars."

He even went up to the ladies' balcony and got a few more contributions: "From a nice lady in the back, ten dollars."

My uncles never publicly announced their donation; that was against their religion. Peer pressure didn't mean borscht to them, and I just assumed that they gave such a small amount because they were not rich, that they were ashamed to announce it. Years later, when I cleaned out their apartment, I found annual receipts for contributions to the First Roumanian-American Congregation for eighteen dollars. In Hebrew, eighteen represents the Hebrew word *chai*, or life, and implies good luck.

The next day, those members of the congregation who were Cohanim recited a special prayer. In ancient times, the Cohanim were the high priests in Jerusalem. The congregants who made this prayer were supposedly descendants of those high priests. As a child, I thought the whole business sounded far-fetched.

During the priestly benediction, the congregation was not supposed to look at the Cohanim up on the *bemah*. You were also forbidden to turn your back on the Aron Kodesh when the ark was opened and the Torah could be seen. And the ark was open during the priestly benediction. This presented a logistical problem. You were forbidden to look at the Cohanim up in the front of the sanctuary when the power of God was allegedly flowing through them to you via the hand signal they gave, which was the same one Mr. Spock made on *Star Trek.* I knew this because I peeked. But you were not supposed to turn around either. Most resolved this dilemma by facing forward and closing their eyes or by using their prayer shawl to cover their heads, and their children or grandchildren's heads, so they could not see. But as in most things, my family operated outside the mainstream. Neither my dad nor my uncles bothered to cover their own heads or, heaven forbid, mine. Instead, superstitiously, Uncle Joe insisted on turning around and facing the back of the sanctuary, directing his backside toward the Aron Kodesh. Every year on Yom Kippur, the Cohanim went up to the *bemah* to give the priestly benediction, Uncle Joe turned around, and the president of the shul ran from his seat on the *bemah* to our row and said to Uncle Harry, "Tell your brother to turn his *tuchas,* rear end, around."

The president didn't bother to speak to Uncle Joe directly. Through the years, he had learned that Uncle Joe was going to do what he wanted. If the Messiah had appeared, Uncle Joe would not have listened. Not that Uncle Harry would give this *k'nocker* the time of day either, let alone act as his messenger. Uncle Harry rolled his eyes and looked another way while the president turned red and stomped back up to his seat. I guess this was why he never approached our row for Kol Nidre donations.

Yom Kippur ended at sunset with the final long blowing of the ram's horn. Sometimes this took awhile. The shofar is not easy to master; it requires a strong set of lungs and plenty of finesse. Mr. Lipshitz, the shofar blower, had the job because he had been a bugler in the Russian army during the Russo-Japanese War of 1905. He owned the store across the street from the shul that sold religious articles. With everyone tired and hungry from a day of fasting, we rejoiced to hear this hunched-over, ninety-year-old man in a long white beard finally sound the concluding shofar blast. If memory serves me correctly, he still had the job when I graduated from high school.

One Yom Kippur, during my high school years, when we walked back to my uncles' apartment, Uncle Harry pointed out the nondescript door of a building on First Avenue between First and Second Street; I could tell it wasn't a residence. But there were no signs up indicating what kind of business was conducted there.

"A joint for the *fagelahs* [gays]," he told me unsolicited, in a matter-of-fact way, no different from the way he pointed out retail businesses on the Lower East Side like Russ and Daughters or Katz's Deli. At the time, I had never given a thought to Uncle Joe's or Uncle Harry's sexuality. I just assumed they were celibate. So I was taken aback when he pointed out the joint for *fagelahs* and wondered how he knew this. Did he notice who entered the place, or supply them with cake, or visit in error, not realizing that the paying guests played on the other team? Or was he gay?

I never asked him, and it never occurred to me to think that a man surrounded by customers who loved him could be so alone.

THE EXCAVATION

1995

People are trapped in history and history is trapped in them.—*James Baldwin*

Within a month of my debut as a lawyer, Nurit, her sister, Cherie, Cherie's husband, Charlie, and I descended on the fifteenth floor of my uncles' apartment building to begin the excavation. Even by New York City standards, it was quite a tangle.

Like my mom decades before, I figured out which three keys on Uncle Harry's key chain matched the three locks of apartment 15G. I pushed the front door open. We advanced into a bouquet of decay. The front door slammed shut, entombing us. My volunteers froze in the foyer. The faint smell of gas got stronger, and I felt my way in darkness through the kitchen to the dinette window. I pulled up the tawny shade, revealing a window so dirty I couldn't tell if it was day or night outside. I forced the window open and turned around as Charlie flicked on a light. We got lucky; nothing exploded.

"Ooooooh myyyyyy God," Cherie said.

Cardboard boxes of various sizes, piled from floor to ceiling, flanked both sides of the front door and continued down the hallway into the living room. The formerly white walls in between the boxes were light gray lower down and turned progressively darker, in bands of neglect, until near the ceiling, where they blackened. A single light bulb, stuck in a socket in the ceiling, stained the living room a weak yellow ochre. A narrow path snaked between the boxes, piles of newspapers, and heaps of trash toward the living room, where the trail ended in a five-foot mound of umbrellas. The white sheets were gone, and a layer of dust now coated everything.

"Mort," Charlie said, "I thought we were cleaning out an apartment. This is a warehouse. What did your parents do? Sublet the place to the Museum of Natural History?"

"No one has lived here for over two years. My parents must have tried to straighten up."

"Oh, yeah? I can imagine what this dump looked like before they straightened up."

People pay to hear my brother-in-law's jokes. And before we were finished plodding through the primordial pathos, his expertise came in handy. He's a psychologist.

We assembled in the kitchen, where the worn linoleum floor was the only place clear enough for the four of us to stand together, albeit in a tight circle. Cherie sniffed near the stove. "There's a gas leak. Mort, you must get it fixed."

"I'll tell my parents; they're in contact with the managing agent."

Cleaning up Uncle Harry's apartment was a job my elderly parents couldn't handle without help. Mom had asked me to pitch in.

Charlie and I followed the trail back into the living room to open more windows. To avoid a cave-in, we squeezed single file around the umbrellas.

"Mort, what's with the umbrellas?"

"When it rained, customers would sometimes forget their umbrellas in the Store. You are looking at the final resting place of seventy years' worth of leftover umbrellas. My uncles never threw anything away."

"Obviously."

The broken venetian blinds could not be raised, so we slipped behind them. Through the soot-covered windows, I couldn't see the terrace I had frequented as a child. We tried to open the windows, but they were caked shut.

"Let's try the bedrooms, Charlie."

We reversed direction and edged down the hallway past the windowless bathroom. Charlie poked his head in. "You don't want to go in there. It's not pretty."

The guest bedroom contained two single beds, night tables that matched neither each other nor the dresser, and an old steamer trunk. Someone had dumped laundered clothes in heaps on each bed but never got around to sorting and folding. After a few minutes of grunting, we forced the windows open. Sunshine and a cold breeze entered, waking and illuminating the dust motes. I slid open the bedroom closet door. The closet was empty. But the closet floor looked chewed up, and I saw what I thought were mouse droppings. They should have kept a Suzy here as well as in the Store.

Years later, a local history buff told me that the story of my uncles reminded her of two brothers named Homer and Langley Collyer, who made headlines in the spring of 1947 when their dead bodies were found in their rat-infested Harlem brownstone buried beneath tons of worthless junk. The rats had gotten to Langley.

As we were leaving the guest bedroom, Charlie noticed the brown leather-trimmed steamer trunk.

"Mort, what do you think is in there?"

"I can't begin to imagine."

Charlie reached down and found the trunk was locked. "My guess, this is where they kept the gold."

Charlie and I inched farther down the hallway to my uncles'

bedroom, past an old-fashioned Hoover carpet sweeper fossilized in between the box-lined walls. In their bedroom, sheetless mattresses rested on matching wooden beds. A huge crack slithered across one headboard. On one bed, two cardboard cake boxes, one filled with pens, the other with rubber bands, were decomposing. On the other bed, a similar box was filled with old belts and rusted buckles. The floor between the set of dressers was covered with buttons, shoehorns, and a fruitcake box. And another huge open box overflowed with a collection of audiocassette tapes. We had found the final resting place of Uncle Harry's bartered tape collection. I opened the window at the south end of the room and looked down Allen Street toward lower Manhattan. The Twin Towers shone bright in the afternoon sun.

We heard our wives open and close kitchen cabinets as they began cataloging, cleaning, and carting. They found stacks of grime-covered dishes, pots, and pans, and Nurit hesitated before touching anything. Her older sister had no such problem. They discussed what to keep and what to throw away.

"Look at these old juicers, Cherie. This metal one is the kind you see in the souk in Jerusalem at the Arab juice-and-nut stands. I wonder how old it is?"

"We need Mom here," Cherie said. "With some of these dishes, I can't tell what is valuable and what is junk. But if someone wants to start a collection of Depression-era glass dishes, this is the place."

Charlie and I returned to the kitchen.

"Mortchela, let me show you something." When my sister-in-law calls me Mortchela, I know the something is not good. Mortchela is short for Mortchela Moishele Katz, who was Charlie's mother's fondly remembered boyfriend in Romania before the Second

World War. It's a nickname designed to soften hard news. Cherie led me to the dinette. Yellowing newspapers, ledger sheets, and notepads still littered the floor.

"Mortchela, you see those holes in the newspapers."

"Yes."

"You have visitors. Mice."

"Then it must be mouse droppings I saw in the guest bedroom closet."

"We'll have to come back several times to clean up this mess. We must be prepared—you can get really sick from mouse droppings. We'll need disposable gloves and masks, heavy-duty garbage bags . . ."

"Cherie, no problem."

"It's a mess, but the apartment has a great floor plan."

I knew Cherie would wear me out. I didn't think it would be fifteen minutes after we started. But I was lucky to have her; she was the best person for this job.

I meandered into the living room. Some choice furnishings hid behind the boxes. A huge off-center framed photograph of Charles Lindbergh in a horse-drawn carriage to the right of the Arc de Triomphe crash-landed on a sleeper couch opposite the watchful gaze of Mr. Mazel, a plastic piggy bank in the form of a chubby rabbi in black coat, yarmulke, and tallis that was perched on top of a 1950s console television. You put the coins in his stomach.

I pulled back a wheeled office chair tucked under a small wooden desk and sat down. I felt the jagged leather edge of the ripped seat through my pants. A brittle piece of brown tape, intended to hold the seat cushion together, fell to the floor. An umbrella poked me in the back. I reached around and pushed it deeper into the pile. Files, papers, and envelopes buried the desk. I opened the desk's middle drawer as far as I could without backing into the umbrella pile. A picture postcard of the Store circa 1980s popped out at me among the stamp pads, boxes of staples, and Scotch tape rolls. I picked it up. Squinting in the poor light, I read the printed top left-hand corner:

Card #6
Ninth Street Bakery
New York, New York

At the bottom the printing continued,

"Sometimes I'd go in there and get a loaf of black bread and go around the corner to the cheese shop and get a pound of sweet cream butter and go home and have dinner."

Roberta Singer

Endangered Spaces, documenting and advocating for New York's cherished cultural landmarks.

I turned the postcard over. Loaves of bread and cake boxes pressed against the picture window like a jigsaw puzzle illuminated by three spotlights. My eyes admired how the image ennobled the Store. But my head said, *lunatics.*

A TALE OF URBAN RENEWAL
1970s–1980s

Life is full of infinite absurdities, which, strangely enough, do not even need to appear plausible, because they are true.—*Luigi Pirandello*

In my freshman year of college, Jacob, my grandfather on my father's side, died. After the funeral service, my dad, my mom, and my first cousin on my father's side of the family, Lew Sweedler, waited in the funeral home's parking lot for the hearse to leave so we could follow it in our car to the cemetery. While we waited, the funeral service long over, Uncle Harry pulled into the lot in his seventeen-year-old Buick. The car had once been rear-ended on the Williamsburg Bridge on a trip back from Pechter's, but it still ran.

Uncle Harry rolled down his window and yelled to us, "Helen, do you want the stuff now?" Before my mom could tell Uncle Harry what he should do with the cake and cookies he had brought to give guests during the week-long shivah period, Lew, who had not seen Uncle Harry since my bar mitzvah, asked me, "Morton, is that your uncle's car?"

His tone revealed his amazement that my uncles were so poor they had to drive around in a car that appeared to have been put through a compactor at the scrap yard. Embarrassed, I hesitated. Before Mom yelled back at Uncle Harry, asking him why he was so late that he had missed my grandfather's funeral service, she responded to my cousin's astonishment without a hint of shame, but with a deep, inexplicable anger.

"It is," she said.

After I graduated from Brooklyn College and embarked on my accounting career in the early 1980s, I rarely saw Uncle Harry and Uncle Joe. I didn't even see them when I slept over at their apartment during the 1980 New York City transit strike so I could walk to my job in Manhattan. I fell asleep in their guest bedroom before they came home and woke after they left in the morning. Meanwhile, Uncle Harry and Uncle Joe remained in the Store on Ninth Street. Nothing had changed there except a photo of a missing little boy named Etan Patz now smiled out from one of the display windows. But Uncle Harry and Uncle Joe were in their seventies, and, as Mom said, they "couldn't do it anymore."

But they did. In protest, intended to motivate them to retire, Dad no longer went to my uncles' apartment on the holidays. I joined him in staying home in Brooklyn. Mom was torn. She still went, not only on the holidays but also to help out in the Store. They held on long enough for one more story.

Late one winter night, Marcelino Casanova walked into the Store. Then again, his name may not have been Marcelino Casanova. I will never know. The only thing I can say for sure is that I am not creative enough to make up a name like Marcelino Casanova.

My mom stood behind the counter wearing her winter coat, hoping Uncle Harry would show up soon so she could go home. Suzy was asleep in the cardboard box on top of the radiator next to Mom. Uncle Joe slept in the back room. No one else was in the Store.

"How can I help you?"

No answer was immediately forthcoming from Marcelino Casanova, except that he took out a gun and pointed it at my mom. I have no idea what he looked like. Mom has no recollection. For some reason I imagine the silent-film star Rudolph Valentino—his hair was slicked back and he was very suave. He said something about handing over all the cash in the register. Although the Store had been robbed many times before, this was a first for Mom. She called for a more experienced hand.

"Joe?"

No answer.

"Joe!"

A groggy "What!"

"There is a man here with a gun."

"Give him some cake."

"Joe!"

Uncle Joe decided his involvement was required and came out of the back room. He quickly read the situation. The pointed gun necessitated more than cake. He opened the register, took out all the cash, and handed it over. The money, the gun, and Marcelino Casanova exited.

The police were called, a description given, and a report filed. Miraculously, the police made an arrest. Uncle Joe received a letter from the district attorney but did not reply.

❦❦❦

April 20, 1983
Mr. Joseph Wolk
350 East 9th Street
New York, New York

Re: People v. Marcelino Casanova
Docket No. 3N025417

Dear Mr. Wolk

Please be advised that the above-captioned criminal case was dismissed by the court because of your refusal to testify even though you were served with a subpoena.

Unless you telephone me within ten days to make a mutually convenient appointment for an interview, this Office will be compelled to discontinue the prosecution of the above complaint.

Sincerely,
PETER HASKEL
Assistant District Attorney

Uncle Joe had no time to answer a subpoena; he was too busy working.

A few years later, my mom and Uncle Joe were both at the counter when Marcelino Casanova walked into the Store. Then again, his name may not have been Marcelino Casanova. The only thing I can say for sure is that he thanked Uncle Joe for not testifying against him, spoke of having a job, and said he was reformed. This time he left the Store with cake and words of encouragement to keep up the good work. No one remembered whether he paid for the cake or if Uncle Joe just gave it to him. We never saw Marcelino Casanova again.

ARTIFACTS

1995

Business and life are like a bank account—you can't take out more than you put in.—*William Feather*

"We're back," Cherie proclaimed to the mice, as I once again swung open the door to apartment 15G. A snowy week had passed since my volunteers and I first experienced the multisensory pleasures of my uncles' old apartment.

Thanks to Cherie, we were armed to do battle with dust, dirt, mildew, mouse droppings, odors, and all things in need of cleansing, even those whose dirt was invisible to the naked eye. For the ultimate word in classifying and valuating the artifacts, a new volunteer was in tow, my mother-in-law, Ellen.

Within minutes, we began the tasks Cherie assigned. Nurit and Ellen were on kitchen duty; I sorted through mounds of boxes to determine what to save and what to trash; and Charlie excavated those sticks of furniture Cherie had decided were worth saving from under the debris of several lifetimes.

I took down a brown rectangular box perched on top of the pile near the front door. The addressee was my grandmother Lena Wolk at 486 Hegeman Avenue, Brooklyn, New York. No zip code was listed. The postmark was from 1957. Although the box was older than I was, it had never been opened. I tore into history with the box cutter Cherie had supplied.

"Warm-o-tray for leisure living, keeps food piping hot, refrigerator-to-oven-to-table, new, delightful! complete with deluxe electric warming cradle, party perfect! Use on ordinary table, buffet, patio. Manufactured by ATLANTIC PRECISION WORKS INC."

Now this was definitely a keeper.

And on it went, box after box, all around the apartment. Royal Chef Thermo-fused Teflon II aluminum cookware, electric carving knives, Corning Ware: the duet set, the bake 'n' fry set, and the galley set. Flatware, glasses, watches, cups, mugs, pots, pans, dishes, transistor radios, plastic service for twelve for those outdoor occasions, clock radios with alarms, clock radios without alarms. All unopened, never used.

Charlie pried loose a standing lamp from the corner of the living room. The lamp protested. A waterfall of the last remains of a yellowing lampshade cascaded over Charlie's brown bomber jacket.

He brushed himself off, looked at me surfing in a sea of shredded cardboard, and asked, "Why did they buy all that stuff if they were never going to use it?"

I stared up at the bare light bulb protruding from the living room ceiling, throwing off its feeble glow.

"They would never buy this stuff if they weren't going to use it. It had to be a freebee."

I retracted the cutting blade and went over to the wooden desk in the corner. As I had last weekend, I pulled open the double bottom drawer. I was remembering the full-page savings-and-loan advertisements in the *New York Post* when I was a kid. *Open your passbook savings account with us and receive your choice of one of the following gifts absolutely free. For an account of only $100 you can receive . . .* The rest of the page was divided according to how much you deposited in the bank. The gifts you could receive were pictured below the corresponding amount.

"Here is your answer, Charlie."

I plunged my hands deeply into the drawer and pulled out its contents over and over again. Bankbooks flowed from my fingertips, reflecting the maelstrom of New York City's ever-changing financial history: American Savings Bank, Anchor Bank, Apple Bank, Bankers Trust, Bowery Savings Bank, Brooklyn Savings Bank, Central Savings Bank, Dime Savings Bank, Dollar Dry Dock Savings Bank, Dry Dock Savings Bank, East Bank for Savings, East River Savings Bank, Emigrant Savings Bank, Empire Savings Bank, First Federal Savings Bank, First National Bank, Franklin Savings Bank, Greater New York Savings Bank, Greenwich Savings Bank, Independent Savings Bank, Jamaica Savings Bank, Lincoln Savings Bank, Manufacturers Trust Company, Metropolitan Savings Bank, Ninth Federal Savings & Loan, Northside Savings Bank, Seaman's Bank for Savings, South Brooklyn Savings Bank, The National City Bank of New York, The New York Bank for Savings, The Williamsburg Savings Bank, Union Dime Savings Bank, Union Square Savings Bank, United Mutual Savings Bank, West Side Federal Savings & Loan.

Multiple accounts existed for each bank. All the accounts were closed—each passbook was perforated with tiny holes to indicate it was no longer in use. Now I knew where the *outside man* must have spent at least some of his time: in banks.

By afternoon, I could no longer completely open the door to the garbage-chute room on the fifteenth floor. It had become the cardboard-box room. I walked down the stairs and dumped the refuse in the fourteenth-floor garbage-chute room. Until it was filled. Then two flights down to the lucky thirteenth floor. The afternoon disappeared.

Images still remain.

Ellen held up a cracked white porcelain cooking pot with a thin blue rim. “Garbage,” she said.

“Mort, what’s the story with the Charles Lindbergh photo?”

“That I want to keep, Charlie. A customer once paid for his Store purchases with that photo during the Depression.”

“He was anti-Semitic, Mort.”

“I don’t care. Don’t ask me why, but I want to keep it.”

The two sisters and their mother stood next to one another at the kitchen sink. One washed, one dried, one directed. Three pairs of bright yellow waterproof gloves moved constantly. Soap bubbles flew. The smell of disinfectant drifted. The 1950s-era wall clock with the solar motif hung motionless in the dinette: this star had burned out long ago.

Charlie called for Cherie to come into the living room, where order was emerging from the chaos. Bank-account bonus gifts lined one side of the room. A telephone bench and a 1930s General Electric wood-veneer combination radio-and-record player silently waited near the front door for a ride to my house. The furniture-to-be-donated pile grew large in front of the still-closed living-room window.

Charlie lifted a wooden end table with a single drawer and leather top.

“Cherie, what pile should I put this in?”

“Now that’s a nice piece of furniture. It looks like the kind of thing you would find in the Kennedy White House. Very classy.”

Charlie put the end table near the front door.

I noticed the old standing lamp near the front door. Its shade had disintegrated; its chord was cracked and frayed beyond repair. The only use I saw for the lamp was to start an electrical fire. I picked it up and started toward the front door, hoping I could shove it horizontally across the remaining air pocket at the top of the garbage-chute room on the thirteenth floor.

"Mortchela, where are you going with the lamp?"

"It's garbage, Cherie."

"No, it isn't," Cherie replied.

"When this thing was new, it wasn't even in the Coolidge White House. I don't want it."

"I do."

I plunked the metal lamp down hard next to the end table. I knew better than to argue with my sister-in-law.

Out of frustration, I pulled open the single drawer of the end table. It was empty except for a booklet I hadn't seen in twenty years: James Madison High School, Ninety-second Commencement Exercises, Brooklyn, New York, held at the Loew's Kings Theater, 25 June 1975. I picked it up, walked over to the dinette table, and sat down.

How did this end up with my uncles? They didn't come to my high school graduation; they were working. I don't remember any sort of acknowledgment from them, certainly no graduation present. I didn't know they knew, or even cared, but on Ninth Street there was silent admiration for my accomplishments.

I thumbed through the pages and found my name:

N.Y.C. English Teacher's Association Award Morton Zachter

I had forgotten. Or perhaps it is more accurate to say that, in the intervening decades, I had no reason to recall I had won an award for being the best English student in the graduating class.

Would my life have been different had I known about the money? Would I have had the courage to ask Uncle Harry for a loan so I could follow my dreams and major in English at a private college? In retrospect, I was sure of two things. Back then the whole business was beyond my control. Now it was time to count my blessings.

CURTAIN CALL
1980s

All things are mortal but the Jew; all other forces pass, but he remains. What is the secret of his immortality?—*Mark Twain*

On a bright summer day in 1984, I met a beautiful young woman in the Amsterdam airport while two armored half-tracks circled the plane I was about to get on.

I can't really say I met her in the airport. I only saw her for the first time in the airport. She wore a sleeveless white cotton shirt, green travel pants, and Israeli sandals. And she was crying. I couldn't imagine why someone so pleasant to look at should be so upset. I tried not to stare, but I did.

In 1981, I began working for a medium-sized CPA firm in Manhattan while getting an MBA in taxation from NYU. Early in my career, I discovered that specializing in taxation interested me more than being an auditor. I would have blown my brains out as an auditor; at least tax involved some creativity, research, and memo writing, and it paid the bills, allowing me to rent an apartment in the East Village two blocks from the Store. Despite the close proximity to my uncles, however, I rarely saw them. For all I noticed, they lived light years away.

I graduated from NYU Business School (formerly the School of Commerce, which Uncle Harry had attended) in June 1984 and then, at my dad's suggestion, had vacationed in Israel. I was heading back to New York to start a new job at the international accounting firm Deloitte, Haskins and Sells. A blond, blue-eyed Dutch beauty who had the looks to be a model sat next to me on the morning flight from Tel Aviv to Amsterdam. I tried to converse with her, but she spoke no English and left the plane in Amsterdam, her final destination.

El Al security policy required that I also get off the airplane in Amsterdam, even though I would take the same plane, and seat, for the second leg of my journey to New York. After a one-hour layover, I reboarded. I neared my seat and saw that the beautiful young woman in the sleeveless white shirt had taken the Dutch woman's seat. I smiled; she smiled back. We introduced ourselves. Despite her Israeli name, she was American. She had led a group of American high school students on a tour of Israel that summer and, like me, was heading home to start a new job. She had arrived in Amsterdam three days before with her friend Didi.

Didi, seated by chance a few rows behind us, seemed pleasant enough, but she was not the beautiful young woman in the sleeveless white shirt. Fate had been kinder to me that day than I even knew. I said very little; I didn't have to, thanks to Stuart. Stuart sat on her other side, and he quickly, and loudly, let us all know that he was a lawyer for IBM. The man had a monopoly on the inappropriate. He bent the beautiful young woman's ear describing in minute detail how sick he had gotten in Israel, throwing up from bad food or tainted water. Stuart was the kind of guy a man dreams about following after he has made his move on a member of the opposite sex. The more he talked, the better my near silence sounded.

I cut in once in a while. I watched her blue green eyes light up when I quoted Mark Twain on the Jews. It was at about this point in the flight that I knew. So did she. This was the person I would marry and start a family with. But I was so shy that I waited almost too long to get her phone number. I asked for it only at the baggage terminal in New York after losing sight of her as we got off the plane. The five minutes it took me to find her again were the longest five minutes of my life. If her bags had come first, she would have been gone forever.

Luckily, my lonely Lark bag bounced down the luggage carousel first. She was impressed that I could travel for three weeks with one small piece of luggage. I helped her lift a carton of books she had bought in Israel from the carousel. The books confirmed my sense of her, and I tucked the paper with her phone number in my shirt pocket and headed for customs.

After I left, she met her mother, who drove her home.

"At the beginning of the flight I was upset," she told her mother. "I wanted to stay in Israel, but I sat next to a lawyer on one side and a CPA on the other. I hope it was okay, but I gave the CPA our number."

"What's the matter with you?" my future mother-in-law replied. "You should have given them both your number!"

Our engagement eight months later started a search for Stuart. We both wanted to invite him to our wedding and seat him next to a forewarned Didi, but we could not track Stuart down. Twenty years later, I am still married to the beautiful young woman in the sleeveless white cotton shirt. As Rabbi Morris Friedman said at our wedding, "It was a match made in heaven."

More amazing was that Uncle Harry closed the Store on the seventeenth day of November in 1985. Only Uncle Harry had attended my bar mitzvah. But Harry and Joe both came to our wedding. It was only the second wedding they had attended since my mom's in 1954. Usually, my uncles never even answered wedding invitations from other relatives, but their behavior was well known in our extended family, and, invariably, the mother of the bride would make a courtesy phone call to my mom, asking politely whether my uncles had received the invitation. In response, Mom mustered all the sarcasm at her command and phoned Uncle Harry. "What's the matter with you? You don't have time to answer a wedding invitation? You're too busy making deliveries?"

I can't recall saying a word to my uncles at our wedding, but they both attended my *ufruf* (a ceremonial calling-up to say the blessing before and after the Torah is read), which Uncle Joe insisted take place at the First Roumanian–American Congregation. At seven in the morning on the Shabbat before my wedding, I returned to Rivington Street and, for what turned out to be the last time, sat next to my uncles to pray. Since the crowd that morning was small, no *machers* were in attendance to comment upon, Uncle Harry told me one of his wedding jokes. Luckily, it wasn't the one about the benefits for married men of separate seating.

"A rabbi in the old country agrees to perform a wedding at another village and hires a driver to take him. When they get to the first hill on their journey, the driver asks the rabbi to step out of the wagon and walk since the old horse cannot pull them both uphill. At the top of the hill, the driver asks the rabbi not to step back into the wagon since the downgrade is steep and difficult for the horse. The rabbi complies countless times on the hilly road. At their destination, the rabbi pays the fare but says, 'I came on this trip to perform a wedding; you came to earn a fare. But tell me, why did the horse come?'"

By 9:00 a.m., services had ended, and a roomful of seventy- and eighty-year-old men washed down their sponge cake with shots of whisky, insisting I do the same. Uncle Joe and Uncle Harry passed on the whisky and left to open the Store.

During our wedding reception, paid for by Nurit's family, my freshly minted sister-in-law, Cherie, approached the bride and told her she should know what she had gotten herself into by marrying me. Cherie nodded in the general direction of Uncle Harry and Uncle Joe, both garbed in vintage suits in desperate need of dry cleaning. It was the last time I ever saw both my uncles together.

"Someday you and Mort are going to have to take care of them financially," Cherie said.

FRUITCAKE

1995

I don't run from jokes about fruitcake. I've changed many people's minds.

—*Mr. John Womble, third-generation owner, Georgia Fruitcake Company*

On our third Saturday clearing my uncles' apartment, Cherie, Nurit, Ellen, Charlie, and I gathered to unveil the steamer trunk, curious what we would find. We had scheduled a doubleheader that weekend and planned to return the next day to finish our work. The sorting of the kitchen and the living room took longer than my sister-in-law had anticipated because my parents came to the apartment during the week and moved Cherie's neatly organized piles around. This was their way of helping.

"Mortchela, big mice are moving things around," Cherie said. "Please call them and tell them to stop. Okay?"

As I lifted the top of the trunk, hinges creaked, and the smell of camphor and mothballs rose. Brittle pages from the July 1966 *New York Times* lined the trunk. That was the month Uncle Harry and Uncle Joe had moved to Manhattan from Hegeman Avenue. And that month, perhaps not coincidentally but in order to avoid packing duty, my mother had sent her *Regards* in a postcard from our Florida vacation. After decades in darkness, a fine white brocade Shabbat tablecloth emerged from the trunk. I carefully removed a fur wrap from another corner of the trunk and ran my hand over the rich fur, past the long tail and rear paws and back again to the mink's face. It had been my grandmother's.

Nurit said, "An old mink wrap. My Tante Regi used to have one just like it."

These objects told me that my grandparents had both the finances and the taste to appreciate the finer things in life: that tablecloth had served for holiday dinners, not food stamps, and there had been fur-wrapped evenings at the Yiddish theater.

From the trunk's very bottom, a crushed manila envelope filled with letters and postcards addressed to my mom on Hegeman Avenue reached daylight for the first time in half a century. In one letter, from the summer of 1945, "the girls from PS 25" invited their coworker and favorite substitute teacher, Helen Wolk, to lunch with them at the St. Moritz opposite Central Park. Despite Uncle Harry's *you belong here* calls, my mom had satisfied her true nature by teaching part-time. Sadly, when I gave her the postcard, along with the contents of the trunk, Mom could not recall going to the St. Moritz. But her mother's tablecloth brought a smile to her face.

"Mort, why is your mother's trunk in your uncles' apartment?" Cherie asked.

"This is the steamer trunk my grandparents used when they came to America in 1913. When my mother married, it stayed at my grandmother's house. When my uncles moved in 1966, it just came with them."

"She should have had this stuff years ago. Your uncles had no business with it."

"It's all hers now, Cherie."

Later, sitting next to the rickety bookshelves in the broken-down brown upholstered chair whose cloth made me itch, I searched for the only book in their library that Uncle Harry had ever showed me: *How to Clean Up at the Race Track.* My uncles' library ran the gamut of Jewish writers from Sholom Aleichem to Leon Uris with a brief pit stop at some early Philip Roth. But in all the years I knew them, even on those lazy High Holiday afternoons home from shul with nothing for me to do but read Roth's *When She Was Good,* neither one of my uncles had ever opened a book. Their library was not for reading; it was just another collection to Uncle Harry. The inside covers of the books, mostly paperbacks, showed the price, usually twenty-five or fifty cents. Uncle Harry had purchased them either from street vendors or, once upon a time in New York, when a never-ending line of secondhand bookstores lined Fourth Avenue from Astor Place to Union Square.

I remember Uncle Harry picking up the book to show me. Its mustard-colored jacketless front cover had just the title, *How to Clean Up at the Racetrack,* printed on it. No author was listed. The binding revealed a different title and an author, however. Uncle Harry handed it to me when I was ten, saying I should study it so I would learn the secret of making it *big.* I opened it. All the pages were cut out in a rectangular shape, and I could see clear through to the inside back cover, where there was a picture of a uniformed man in a cap holding a broom and pulling a metal garbage can on wheels with a shovel hanging from it. At the bottom end of the broom handle, little pieces of straw were attached to the back cover. Uncle Harry got a huge kick out of showing me that book. I thought it was a stupid reason to destroy a book; my mom had taught me to treat books with great care. But I ran my finger over the coarse straw bristles as I nodded to show Uncle Harry I understood his message. It was his way of telling me that working hard was the only way to make it in this world. I never found the book.

But I did find Mrs. Rose Weinstein as she lay among the detritus of my uncles' lives. I uncovered her on a LIFO (Last-In-First-Out) basis. The top layer of papers in one desk drawer was of recent vintage. But as I dug down deeper through the decades, I uncovered Mrs. Weinstein in a mid-1960s stratum. I uncovered, at least, the paper trail that linked her to Uncle Harry. When I read her name on a paid-in-full receipt from December 1966 for forty-one dollars rent for an apartment at 235 East Second Street, I heard my mom thirty years before.

"Where are you running? To Weinstein?" she would say to Uncle Harry, pronouncing the *W* as a *V*, more as an accusation than as a question.

As a kid, I had gathered from bits and pieces of conversation that Mrs. Rose Weinstein was a bedridden woman in her seventies or eighties without family and that Uncle Harry took care of her. I never understood her connection to my family, and, as usual, I never asked.

After a while, her name was no longer mentioned, and I assumed she had died. She never crossed my mind again until I hit the 1966 stratum and unearthed a mother lode of bills paid on her behalf by Uncle Harry: not just the rent, but Con Ed bills and medical bills. A bill from a Peter Kir, M.D., for $216, not an insignificant amount back then, indicated that he had made two (sometimes three) house calls a day to Mrs. Weinstein during the last two weeks of October 1966 for "feeding" her, "dressing of both legs with antibiotic ointments," and administering various medications. The doctor even left a list of foods he suggested she could eat. Why did Uncle Harry take such good care of Mrs. Weinstein? Who was she to him?

A few weeks after I unearthed the bills, my mom cried when I asked her about Mrs. Weinstein and gave the phone to my dad. He said there was no connection between Uncle Harry and Mrs. Weinstein other than that she was a regular customer at the Store, was poor, and had no one. Uncle Harry felt sympathy for her.

"But he paid her bills, Dad. Her Con Ed bill was even mailed to the Store."

"She didn't understand finances and trusted Uncle Harry."

"Did you ever go up and see her in the apartment?"

"No. The few times I drove Harry over to Second Street, he went up by himself. I waited downstairs in the car. He had a key."

He had a key. That is a high level of trust to give a man who sells you day-old bread. A mixed baker's dozen was my Uncle Harry. He never paid my mom a dime for a lifetime of toil, but he gave both his time and money for a customer. Could Uncle Harry's charity radiate to a nonrelative as though he were a male version of Mother Theresa? I always thought Uncle Harry did nothing but work. Now I was not so sure.

I found a photo of Uncle Harry in the 1940s looking dapper but bored in a double-breasted suit. He was sitting at a table in a bar with a smiling, dark-haired, well-dressed woman wearing a silk scarf. In between them sat a sailor; behind them, at another table, a woman sat on a man's lap. In another photo, the same dark-haired woman appears with a different man. When I showed the photos to my mom, she had no idea who the woman or the men were, but she told me that, as far as she knew, Uncle Harry never had a girlfriend.

So how did Uncle Harry's relationship with a woman a generation older than he fit into this story? Was she really only a customer? Was Mrs. Weinstein somehow connected to Max or Lena? An enigma in life, in death a man forever lost in the city's penumbra, Uncle Harry lived a loveless, uncomforted life punctuated by what appeared to be a singular act of great kindness.

I sat lost in thought, instead of cleaning out my uncle's bedroom. Everyone else was working. With memories of *How to Clean Up* fresh in my mind, I felt like Tom Sawyer whitewashing the fence with his friends. I walked through the dinette to the kitchen. Next to the wall-mounted phone in the dinette, a piece of gray cardboard hung taped to the wall. In thick black crayon, it read MORTON 516-569-7694. After Uncle Joe died, when Uncle Harry was lonely or had an emergency and could not reach my parents, I was his next call. In all the years I knew him, he never once called me.

Down on her knees, Cherie was cleaning out old jars from underneath the sink. I had entered just as she removed an ancient jar of borscht. Wine ages; borscht turns green. Cherie could not figure out how to dispose of moldy borscht; if she opened it, to spill the congealed mass into the sink, we might all perish from the stench. But she didn't want to recycle the glass without emptying it first. Cherie put the jar down on the counter. Next, she picked up a little black frying pan I had not seen since Lindsay was mayor of New York City.

"Do you want this frying pan? It's still usable."

"No, Cherie, Uncle Harry used to make me scrambled eggs in that pan. Like he used to make in the goulash joint."

"What is a goulash joint?"

"I'm still not sure. You can throw it away."

"It sounds to me like you'll wish you had kept this. It's a memory."

"That's why you can throw it away."

As usual, my sister-in-law was right. I wish now I'd kept that frying pan.

I left Cherie to figure out the best way to dispose of twenty-year-old borscht and headed back to my uncles' bedroom. There I unearthed a small box with a stenciled imprint that read *Rum and Brandy Flavored Fruit Cake.* My uncles sold a lot of these in their day. I wondered what was inside; judging from the buttons next to it, I figured it was a sewing box.

I sat down on one of the beds and examined the rectangular box. It was designed to look like wood but was clearly cardboard. One end of the box read, "Rum and brandy flavored fruit cake is the result of a time-tested formula, artfully blended, with choice fruits and nuts, slow baked to give you a delicious cake, which mellows with age. Spilke's Bakery of Brooklyn."

I opened the box. Thick circular rolls of paper held together with thinly stretched rubber bands filled the box. I picked one up. The rubber band fell apart when I touched it, a portion stuck to the outside of the roll from heat or age or both. The roll unraveled in my hand in slow motion, and bits of brittle paper fell between my fingers to the floor like a fine green waterfall.

The outside layer of the roll had once been a one-dollar bill. I gingerly lifted the corners of the bills; the denominations got higher and higher. But Uncle Harry must have neglected to clean the cake boxes before placing the money inside them. Most bills were fused together and could not be separated. Better preserved deeper in the roll, the innermost bill was a fifty.

I rummaged through the rubble for more fruitcake boxes. One sat atop the other dresser. Hidden behind that dresser were more fruitcake boxes, one holding nothing but two-dollar bills. I'd never seen a two-dollar bill before and was sure they were fake. I yelled for Charlie. After his initial shock,

Charlie explained that two-dollar bills were once issued, but no one used them. Eventually the Treasury stopped printing them, and they disappeared from circulation. These two-dollar bills were indeed real. All the money was real. But now they were just pieces of paper, falling apart or stuck together but mostly worthless. The salvageable bills amounted to a few hundred dollars. What a waste. Typical, I thought. After all I had learned in the past few months, I was not excited by crumbling two-dollar bills.

"Mort," Charlie concluded, "you're from old money."

We alerted the rest of the crew to be on the lookout for fruitcake boxes. My mother-in-law said my parents had brought her a fruitcake box the first time they ever met. It had a real fruitcake in it. *Just my luck*, she said. She still has it in her freezer as a memento.

After this, I found canvas bags in my uncles' bedroom closet. Eggshell white and closed at the top with a drawstring, they were the kind you see Brinks men carrying into banks. The pennies, nickels, and dimes my uncles had used to make change bulged within the bags. Bags of small coins don't add up to much money, I learned, but they do weigh a ton.

Charlie and I carried the bags to the elevators, one bag in each hand, our arms weighted down, and the bags almost touching the floor. I wondered what people would think if they saw the two of us schlepping bags obviously filled with money. The elevator didn't stop on the way down to the first floor. *Good*, I thought, *we can slink away unnoticed.*

My father-in-law, Daniel Roshin, waited in his car in the best parking spot, directly in front of the building. We did not want to leave the car alone as we filled it, so Danny was our watchman. We had the right man for the job. Daniel had worked at Manufacturers Hanover letting customers in and out of the vault.

Under cover of darkness, Nurit and I returned to our home in Woodmere, Shabbat long over. This pleased me; no one walking back from shul would see us as we carried our bounty into the house. Inside, Nurit's uncle and aunt waited along with Cherie and Charlie's fourteen-year-old son, Eli, who was quite impressed by the sacks of coins and the fruitcake boxes. "Can I go with you tomorrow? Please let me go back tomorrow!"

Based on the stories he had heard and the numismatic archeological findings, Eli must have thought my uncles' apartment was a cave filled with buried treasure.

"No. You have a math test Monday. You have to study," Charlie said.

"Please, I must go."

"No!"

"Pleeeeeeeeeease. I could study after we come back. I must go. I fit into small places and could find fruitcake you guys wouldn't even know is there."

On Sunday morning, Eli joined us on our trip to the apartment. I figured we had found all the canvas sacks and fruitcake boxes the day before, so I told Eli if he found any he could keep them. Several times that afternoon, Eli bellowed out from my uncles' bedroom, as if he were in the crow's nest of the *Pequod* and had sighted Moby Dick.

"Uncle Morty, more fruitcake."

On Monday, Eli failed his math test. It didn't matter. A few years later, he attended Dartmouth, where he grew to appreciate the school colors, one of which is a deeper and more natural shade of green than he had seen in my uncles' apartment. There he excelled in subjects of his own choosing. At his graduation,

I cheered when they announced his name at the Commons in Hanover, New Hampshire. Nurit and I made sure to give Eli a generous graduation present. On that day, I lived through him a part of life I had missed. But by then, enough time had passed for me to have mellowed, like a well-aged fruitcake.

SINGING IN THE RAIN

1991

Look toward heaven and count the stars, if you are able to count them. So shall your offspring be.—*God to Abraham, Genesis 15:5*

My toes squished about in my soaking wet wing tips as I sank into the backseat of a New York City cab on my lunch hour. A late February rain pounded the cab. The midtown traffic was ferocious. I was going nowhere.

I stared at my shoes. When I was in college, I had promised myself I would never wear wing tips. Back then, I was sure only Republicans wore wing tips, not liberal Jews from Brooklyn. But adulthood has a way of eviscerating expectations you once took for granted. When I had worked at Deloitte, on the 101st floor of One World Trade Center, I learned that the key to success in the business world was to "dress British, think Yiddish." Since then, the only shoes I wore to work were wing tips—one brown pair and, to confirm my identity as a thirty-three-year-old accountant, the black ones I was wearing today.

It was 1991. The cabby and I sat on the north side of Forty-second Street in front of Grand Central Station, heading west. I was breathing hard, sucking in the odor of my damp charcoal gray pin-striped wool suit. My feet felt so cold. The driver hadn't taken his foot off the brake in five minutes—an eternity for a New York City cabby. For me it was a lifetime, and it wasn't even my lifetime.

The cabby mumbled in Arabic; I mumbled in English. We were not mumbling to each other, just to ourselves. It reminded me of Lily Tomlin's joke that all the homeless people in New York City should be paired up. That way, when they talked to themselves, at least it would look like they were having a conversation.

The radio played softly. The disc jockey said it was Carole King, but I knew better. The woman singing, whose real name was Carole Klein, was my fellow alum of the County of Kings . . . *It's too late . . . we really did try . . . Somethin' inside has died . . .*

The cabby had found me, with upraised arm, delirious on York Avenue in front of New York–Cornell Medical Center. The specialist I had seen there was *the top man* in his field. My doctor was the best. My doctor was the brightest. My doctor was on the cutting edge. My doctor wrote the book. My doctor couldn't do shit for me. But on that rainy day, he had tried to send me off with a glimmer of hope.

His last words to me were, "In ten years, I guarantee you, technology will have advanced to a point where something can be done."

I responded with a polite *thank you* and got up to pay his bill. I didn't bother to tell him what I really felt. I didn't have to; he knew just by looking at me. My face wore defeat. Even if I qualified for the procedure he had suggested, our medical insurance did not cover the ten-thousand-dollars-a-pop treatments. We could take out another mortgage on our home to pay for the treatments. But if they didn't work, and we needed funds to adopt later, what would we do? A faint, distant star was just not good enough. But that was all he had to offer. I silently screamed, awash in an agony so great I wished I were dead.

I have a good wife sitting on schpilkes *in her office waiting for the phone to ring. She wants to hear that a man who got the chicken pox four years ago, at the age of twenty-nine, can father her children before he turns totally gray. She doesn't want to hear about what might be in ten years. And I don't want to give her that news—between my bachelor uncles and me, firing blanks, I am the last of the Mohicans. I have done everything you asked—masturbated into beakers in your sterile offices more times than I care to recall, had intercourse on cue at the right time of the month for years until it was anything but making love, had a vericocelectomy. You held the scalpel. My fellow Jew, my doctor, you know what the Torah commands, be fruitful and multiply. Please, help me.*

I remained submerged in grief as we crawled down Forty-second Street between Sixth Avenue and Broadway. The light turned red. I told the cabby I wanted to get out, but he couldn't pull over. The traffic was bumper to bumper. I couldn't wait any longer. I paid him, flung open the yellow cab door, and swung my spongy wing tips right into the flooded street. The sky was ash. The only light I could see was the never-ending news ticker of One Times Square; and the news was not good—Saddam and Bush and the first Gulf War. Head down, I walked to Forty-third Street and entered 1500 Broadway. I shook off my umbrella, straightened my tie, ran my hand through my hair, dabbed my eyes with the crumbling ball of tissues I pulled out of my back trouser pocket, put on my work face, and took the elevator up to my job.

Just another New Yorker, back late from lunch.

ANCESTORS

1995

I don't know who my grandfather was, I am much more concerned to know what his grandson will be.—*Abraham Lincoln*

I sank into the upholstered chair with broken springs in the corner of my uncles' living room. Charlie sprinted over; I hadn't seen him that excited since we found the fruitcake boxes.

"Did you see these?" he asked.

Like a poker player laying down a royal flush, Charlie revealed a collection of card-sized, black-and-white photos of scantily clad women from the 1920s. "You know how much these must be worth?"

Embarrassed, I extricated myself from the depths of the chair and stood. The odor of disintegrating stuffing arose from the musty cushions with me. In the thirty years that had passed since my bachelor uncles moved into this apartment on First Avenue between Second and Third Street, dusting had been an afterthought. I grabbed the photos from him, and after a quick look, I stuffed them in my back pocket. "I have no idea," I replied. "My father told me he had found some *funny stuff* in the apartment. I guess this is what he meant."

We had reached our last day of cleaning the apartment. I had saved the books and paperwork for the very end. What we did not take home that day would be left in the garbage-chute room for the tenants of the fifteenth floor of 40 First Avenue. Although, at the time, they were of little use to me, we kept all of Uncle Joe's prayer books with God's name in them; you cannot throw them away according to Jewish law. But how many paperback copies of Chaim Potok's *The Chosen* does one need? My uncles had three, and despite recent upheavals that had left me feeling chosen, I kept just one.

After I had boxed the books I wanted, I sat down and thumbed through a true treasure, *Greatest Jewish Sports Heroes,* with a section about Sandy Koufax, the Hall of Fame Dodger pitcher. I never saw him pitch in Brooklyn, where I was born four months, thirteen days, and nine hours after the Dodgers played their last game in Ebbets Field on September 24, 1957. The Brooklyn Dodgers and I passed like two ships in the night, symbolic of the ineffable I was close to but never actually knew.

It didn't take me long to thumb through the book; there are not that many Jewish sports heroes. Soon I was back where I had started weeks before, sorting through piles of papers at the desk beneath the window whose broken blinds still refused to open and let in the dusty sunlight. There I determined what I wanted to keep or try to forget. Behind me, the umbrella pile had melted and disappeared. In the empty space were new white file boxes filled with old papers I thought worth saving. Cherie, as always, had thought ahead, anticipated any and all needs, and brought the boxes for just this purpose. Black heavy-duty garbage bags stuffed with yellowing electric bills, stale checks payable to the Ninth Street Bakery, and Depression-era checks payable to Consolidated Edison signed, in a clear bold script, by my grandfather Max Wolk, lined the walls.

Then, in the muck I found a file with several pieces of paper headed SAFETY FIRST GUARANTEE. They read as follows:

THIS CERTIFIES that I, the undersigned, a female about to enjoy sexual intercourse with______________ am above the age of consent, am in my right mind and not under the influence of any drug or narcotic. Neither does he have to use any force, threats or promises to influence me. I am in no fear of him whatever; do not expect or want to marry him; don't know whether he is married or not and don't care. I am not asleep or drunk, and am entering into this relation with him because I love it and want it as much as he does, and if I receive that satisfaction I expect, am willing to make an early return engagement.

FURTHERMORE, I agree never to appear against him, or to prosecute under the Mann White-Slave Act.

Signed before going to bed, this _____day of ______193__

By ______________________

Address ______________________

Witnessed by: ______________________

Now, you don't find something like that every day. I wondered who had used them. The file contained a half-dozen waivers, some dated, one even signed, but not by anyone in my family. In law school, I had learned that the Mann Act was an old federal law that made the transportation of women across state lines for purposes of prostitution illegal and punishable with jail time.

I showed the waivers to my psychologist brother-in-law. Charlie perused them. Then, like any outstanding member of his profession, his thoughts turned immediately to sex and money.

"How do you start a business like that?" he asked.

"These are too old to be Uncle Harry's. Do you think my grandfather was a pimp?"

"Maybe. I don't think you will ever know for sure. What do you know about your grandfather?"

"Not much. My mother told me he was tall, smoked, had such a good memory he never wrote anything down, and once sold a formula for yeast to some company for a lot of money. That always sounded fishy to me. But let me show you something else."

I pulled out an old bank book I had found for the account of Max Wolk with the United States Bank located on Delancey Street. Handwritten deposits dating back to the 1920s lined the pages. I handed it to Charlie.

"Look at the deposit on January 16, 1930."

"It looks like it's for $9,105. It's hard to tell. Almost all the other individual deposits are for amounts less than $100."

"Charlie, do you have any idea how much money $9,000 in 1930 would be today? Well over $100,000, adjusted for inflation, maybe more. Where did my grandfather get $100,000 four months after the stock market crash of 1929? You don't accumulate that kind of money from a commissioned bakery with penny margins."

"Maybe it's from the yeast sale. Or he saved, and only made the deposit because he was afraid to have that much cash around."

"You're right. I'll never know for sure. But I intend to try to find out. I can't ask my mother. What would I say? Mom, by the way, could you tell me if grandpa used the bakery as a front to launder money from an illegal business? Prostitution perhaps? Or did he just like to frequent hookers or collect unusual pornographic material?"

Charlie handed me a yellowing, wrinkled, and stained piece of paper he had found: folded in quarters, ripped down the middle, held together by only a thin strand of Scotch tape. It was a copy of the *Desiderata*, reportedly written in 1692 and later found in old St. Paul's Church in Baltimore. At the very bottom it was signed, Max Ehrmann 1927. Judging from the fact that it was Scotch taped, it was important to someone. To whom, I wondered? And who was Max Ehrmann? Because the signature was 1927, I assumed it also had belonged to my grandfather. I read it and gave it back to Charlie. Despite my spiritual black hole, I asked him to put it in a file in one of the white boxes with the papers I intended to keep. Years later, I learned who Max Ehrmann was, and that a desideratum is something desired or essential. Even today, when I think of my uncles and the grandfather I never met, of all the papers and objects we found in my uncles' apartment, the *Desiderata* is the one I choose to remember.

Alone at day's end, I walked through the apartment one last time, not to relive old memories but to make sure I had taken everything I wished to keep. After weeks of sorting, all that remained were a few sticks of furniture waiting for the Vietnam Vets to pick up. The squeaking of my shoes on the linoleum floor Cherie had insisted on washing echoed in the empty apartment. I double-checked the living room to make sure I had not left *Greatest Jewish Sports Heroes* on the stuffed chair. I hadn't. After I locked all three locks and went down the elevator of my uncles' apartment building for the last time, I remembered that I had not opened the terrace door to check if there was anything worth saving outside on the terrace. Come on, I thought, as I recalled the one thing my uncles kept there thirty years before. Who was I kidding?

Downstairs, Charlie's Land Rover was packed and loaded, but the 1920s photos were not in the truck. I didn't want my mom ever to see those photos, which, from their vintage, I assumed could have belonged only to my grandfather, the man I am named for. I had thrown the whole business down the garbage chute. To this day, my brother-in-law still tells me how stupid I was to throw those photos away.

"Mort, you have no idea how much those pictures would bring on eBay. No idea."

Even without the photos, after weeks of culling, an abundant collection crammed Charlie's truck: a picture postcard of the Store, hundreds of postcards mailed from all around the world, a Manischewitz Haggadah, and correspondence from the Manhattan attorney-general's office involving a criminal case.

THE CALL
1994

Just when you think you've graduated from the school of experience, someone thinks up a new course.—*Mary Waldrip*

In June 1994, I celebrated graduating from Brooklyn Law School by upholding family tradition. I didn't attend my graduation either. I wouldn't have attended even if subpoenaed. I worked that day as an independent contractor for a Manhattan law firm with a niche practice in estates and trusts. I did so to gain experience so I could pretend I knew what I was doing when I started my own legal practice. I also needed the dough. The firm paid me ten dollars an hour, and I couldn't afford to give up the day's wages. At the graduation ceremony, my name was called as the recipient of an award as the best student of taxation in the graduating class, but I wasn't there to accept.

I took the New York and New Jersey bar exams the last week of July. The next week, it was back to work. But freed from the time law school had taken, I could dedicate myself exclusively to expanding my practice. I had no choice; we could not get by for much longer on my current billings. As always, my prime motivation was anxiety born of fear of failure—in this case, failure to provide.

One evening that week, I received a hysterical call from my mom.

Your father, colon cancer, surgery, yesterday, successful, stable, hospital.

The next morning, Mom and I entered Dad's hospital room. He lay in bed, the midday sun shining directly on his face, revealing clear eyes and remarkably good skin tone. He started chirping at us in a surprisingly strong voice as soon as we walked in. "I'm all right. I'm all right. They got it early before it had a chance to spread."

Mom and I sat at the foot of his bed, and we talked for hours. Apparently, they had discovered the cancer early during a regularly scheduled colonoscopy.

"From now on," I said to my dad, "you must let me know what's happening. Tell me ahead of time if you are going into the hospital for surgery. I can only help if I know there's a problem. No more secrets, okay?"

"You were taking the bar exam and had enough on your mind."

I remembered seeing a television interview from Beijing during the Tiananmen Square uprising. The camera showed two elderly Chinese grandparents. The man could not get out of bed; the woman could barely walk. They lived in a multiple-floor walk-up. The city had exploded in violence all around them. They kept pleading to the camera, to their children and grandchildren in America, "Don't worry about us, we will be okay. There is nothing to worry about."

After I pestered my dad some more, he agreed I was right.

"In the future," he said, "we will call you in an emergency."

This was a bigger breakthrough than mainland China surrendering to Taiwan and turning democratic.

Then Dad leaned back on his pillow and lifted his chin. "Go to New York and empty out the box."

By New York, I knew he meant Manhattan. "What box?"

"Uncle Harry's post-office box."

"What does Uncle Harry need a post-office box for?"

Mom screamed at me, "Don't ask so many questions!"

"He gets a lot of mail," my father said.

He told me Mom would give me the keys and that I should also check out the mailbox in the lobby of his apartment.

"And while you're there, go upstairs and see what's doing in the apartment."

What could be doing in that apartment? I figured the roaches must be having a field day. Mom and Dad kept paying rent on the apartment to sustain Uncle Harry's fantasy that he might someday return there, despite his worsening dementia.

"I'll go tomorrow afternoon. We'll come to visit you in the morning."

"Now go with your mother and get a bite to eat in the hospital cafeteria. You both must be hungry. The food is excellent."

Great. Just what I wanted for lunch—hospital food. But to my parents, this was a rare opportunity. The cafeteria was kosher. Here they could eat meat at what was in their minds a kosher restaurant, at subsidized prices. I followed orders and took my mom to the cafeteria.

After lunch, I drove Mom back to my parents' apartment. On Bedford Avenue, we passed a huge house that sprawled over four lots—two on Bedford Avenue, and two directly behind them on Twenty-sixth Street. Four houses had been ripped down to build this one three-story house, which dwarfed every other home on the block. It was surrounded by a high brick fence with steel posts. The edifice screamed: Look at me, I'm rich—but you keep your distance outside my massive walls.

I made the left onto Kings Highway and got lucky. I found a parking spot right in front of my parents' tenement. Alternate-side-of-the-street parking had ended at 2:00 p.m., and most people had not yet returned from work to grab the choice spots. I even got to parallel park. For a city boy lost in suburbia, somewhere between the mall and my driveway, that was the highlight of my day.

The graceful sixteen-foot-high lobby of my youth, once filled with furniture—cream-colored leather chairs and sofas and a leather bench encircling a large pillar in the center of the room—was now barren. The lobby's sole piece of furniture was an unfinished wooden bench built around the pillar. The high, majestic windows, with their stained-glass panels of medieval knights and castles that had charmed me as a little boy, were broken, replaced with regular glass or cardboard. Only a few stained-glass panels remained. A garish yellow paint sullied the once-proud lobby walls.

Inside my parents' apartment, no lights were on; all the shades were drawn; and it was hot as hell. But the apartment was not vacant. Uncle Harry sat in the living room, his feet resting on a hole in the once green Karastan carpet. He floated on pillows and newspapers tucked beneath him, and I feared if I removed those he might plunge clear through the seat and worn-out carpet down to the first floor of the building.

"Hello, Uncle Harry."

No response came, only the dull gaze of bewilderment.

"It's Morton," Mom yelled at him. "Harry, remember Morton?"

I was surprised at how far Uncle Harry had deteriorated. He nodded, his eyes vacant, without a spark of recognition. I held his arm while I helped him up to go eat lunch. His whole body had shrunk. He disappeared inside a long-sleeved white shirt that was too big for him. I remembered how he could lift crate loads of merchandise from the trunk of his car. Now he shuffled across the carpet and the linoleum floor to the foyer to the dining table. The linoleum was so worn that in spots the wooden subfloor showed through.

After a few minutes of frantic frying, Mom put together a plate for Uncle Harry—a tuna-fish cake with some lettuce and tomato and a piece of Wonder Bread. When I was a child, the only bread I remembered Uncle Harry eating was fresh bread from the Store. How the mighty had fallen. If Uncle Harry had known he was eating prepackaged Wonder Bread, he would have keeled over and died from shame right then and there. I sat down across from him. Uncle Harry cut the fish cake with his fork, picked it up with his sharp knife, and started to bring it to his mouth. His hand shook from Parkinson's. I reached out and yelled for him to stop, but he just kept going. Mom grabbed the knife out of his hand, cut up his food for him, and left him with only the fork. There was no conversing with Uncle Harry. It didn't matter what topic I brought up. The man who had always had a joke to tell now said *okay* and nothing more. I couldn't even tell what he was saying okay in response to.

It was a sweltering August afternoon. The clatter of jackhammers blasted through the open dinette window. Distracted, I got up and sat down on the fourth matching vinyl chair from the dining-room set. It served as the telephone chair and was located in the hallway next to a wire stand covered with cardboard, phone numbers scribbled all over it. From above my head, a river of stains ran down the walls from the ceiling. When I had lived here as I child, sleeping in the dinette with my head next to the Frigidaire, the upstairs apartment bathroom had leaked. Some things never change.

But some do.

THE LAMP

1995–2001

I returned and saw under the sun, that the race is not to the swift, nor the battle to the strong, neither yet riches to men of understanding, nor yet favor to men of skill; but time and chance happen to them all.—*Ecclesiastes 9:11*

Old friends said, "It couldn't have happened to a nicer guy."

My brother-in-law said, "Since it wasn't me, it might as well be you." Charlie expressed what everyone who knows my story feels. But I was my uncles' nephew: although my windfall allowed me the opportunity to take few new clients, I kept working as a CPA and tax attorney.

For years since my first act as a lawyer, I had wanted to give up my tax practice and focus on writing full-time, but something held me back. After all those years of struggle finishing evening law school, how could I give up my practice? I knew myself. I was not the kind of person who could write and practice law at the same time. I'm no Scott Turow.

But I wondered how I could give up a practice I had worked so hard to establish for a writing career in which I'd probably never make a dime. What would I tell people I did for a living? Write? How could I say this if I had never published anything and perhaps never would?

Yet we had more money than we would ever need. My mother's singular act of kindness in disclaiming the bulk of Uncle Harry's assets meant that continuing to work at a career I had never wanted was absurd. But despite feeling unfulfilled, that was exactly what I did. At some point though, it had finally dawned on me that I was as crazy as my uncles, my exemplars of how to hoard money after accumulating millions. But I still couldn't change. Every month, I added up the totals from the brokerage statements and counted my millions, getting a perverse thrill seeing the balances go up, not so different, I imagined, from Silas Marner and his coins, or from Uncle Harry.

Six years dragged on this way. I found excuse after excuse not to abandon my tax practice, holding on to it like a crutch, denying myself.

The summer of 2001 evaporated uneventfully. The big news story that summer was whether a California congressman killed his young intern who had disappeared. We didn't know how good we had it. Fall had no end.

As I drove north on the New Jersey Turnpike past the Meadowlands one Sunday in October 2001, I forced myself to look east, across the Hudson to my beloved New York, to see again what hate had done. Seated in the passenger seat next to me, Nurit asked, "Mort, where were they?"

"Downtown, in the void beyond the World Financial Center. See the empty sky?"

If I had stayed at Deloitte, and if they had not moved across the street to the World Financial Center, I could have been working there that morning. It could have been me on the 101st floor instead of the employees of Cantor, Fitzgerald.

It was time to stop feeling guilty, time to stop worrying about what others would think, and time to break my family's pattern of insanity. It was time to use the dough to live my dream.

Although I did not know it then, Joe Temeczko, the Minneapolis scrap collector, died that October. He had been so profoundly affected by September's disaster that he changed his will and left $1.4 million to the City of New York, which was used to plant flowers all across the city as a living memorial to the victims of 9/11.

That night, after my kids were asleep, I went into my study, turned on the ancient standing lamp I had wanted to throw away, but which Cherie refurbished for me anyway, took out the Mont Blanc pen Charlie once surprised me with for my birthday, conjured up what I had learned in the writing class Nurit had signed me up for, finally took advantage of the gift of time my mom had given me, and started to write this book.

CODA
2002–2006

Nothing is allowed to die in a society of storytelling people.—*Harry Crews*

I opened the faded front door of the First Roumanian–American Congregation one rainy evening in the fall of 2002. The synagogue was now closed most of the time, and I arranged my visit to coincide with evening services. I climbed the creaky stairs to the second floor, past the dusty marble donation plaques still hanging in the hallway, and into the main sanctuary, where the water-stained ceiling reeked of mildew. The congregation had dwindled and was barely holding on. The main sanctuary was dark and unused. The small room downstairs was all they needed, even for the High Holidays. I walked past our old row to the front of the empty sanctuary, half-expecting to see a dead bird on the floor beneath the windows.

After a few minutes, I went back downstairs to make sure they had a minyan, a quorum of at least ten men needed to say kaddish, the memorial prayer for the dead that concludes evening services. I was the tenth man. I prayed and thought of the Store, of the lives wasted there, and of the lost world of the Lower East Side.

A few years later, the water-damaged roof of the First Roumanian–American Congregation collapsed. The congregation lacked sufficient funds to restore it, and in March 2006, the 150-year-old building was razed.

After my family moved to New Jersey, we joined Beth El Synagogue in East Windsor. This synagogue also had a leaky roof and badly needed refurbishing. Nurit and I became involved in the campaign to renovate the synagogue, and a new sanctuary was built. In doing so, I hope we taught our children the meaning of the word charity—*tzedakah.*

Like my uncles, the *Desiderata* is not what it appears to be. In the 1950s, a rector at St. Paul's Church in Baltimore read the *Desiderata* as part of his sermon and reproduced it on the church's stationery, whose letterhead indicated that the church was founded in 1692. This started the myth that the *Desiderata* was written in the seventeenth century. It wasn't. The *Desiderata* wasn't my grandfather's; it couldn't have been. My grandfather died in 1937, and the copy from my uncles' apartment did not exist before the rector's speech in the 1950s. The *Desiderata* belonged to Uncle Harry; and based on my other finds such as blue-joke books, French decks, dirty-joke cards, a bronze tie clasp showing a man's hands holding female breasts, a business card for late-night barefoot dance parties on Sixteenth Street off Union Square, a metal token good for an all-night stay at Stella's Saloon in Virginia City, Nevada, featuring *the best screw in town,* I'm now sure the Mann Act waivers were his as well. In the end, I have reached the conclusion some people collect coins or stamps; my lonely uncle preferred humorous pornography.

Neither Uncle Harry nor my grandfather had a criminal record. I checked. If they had ever done anything illegal, they never got caught with their pants down. The Outside Man, forever flashing a smile and telling a joke, was just out getting laid in between stops at his banks, at Pechter's, at Mrs. Weinstein's, or at the Second Avenue Deli. No crime in that. Where and with whom? Does it really matter that I will never know? But when I now read Hassan's innocent babka business letter, I sense that Uncle Harry's "friend at the Candy Factory" was a close one.

What is clear is that Uncle Joe, the perpetually dour Inside Man, didn't get out enough to pray, let alone get laid, due to Uncle Harry's selfishness. And thanks to Uncle Harry, my mom received a whole different kind of shafting.

My mom and dad still resist taking car service anywhere, but they do spend Passover at my house, along with Ellen, Cherie, Charlie, Nurit's uncle and aunt, Eli, numerous cousins, and their children. We say kiddush, drink lots of wine, use numerous Haggadahs, argue, and sing. The children take turns asking the Four Questions and even get some answers. I pay the children a king's ransom to get back the *afikoman.* I don't let my mom near the kitchen. No one has to count food stamps. For this, and for much else, I count my lucky stars every day.

My grandfather was not so lucky. After consulting with a handwriting expert, I learned that my grandfather's deposit with the Bank of the United States was for $91.05 not $9,105. The Bank of the United States, which was used by many Russian immigrants, went bankrupt on December 11, 1930, two days after my grandfather made a deposit of $55. My grandparents lost everything in the account, a total of $7,550.05. It was probably their life savings from, surprisingly, the sale of dough.

A 1918 postcard I found in my uncles' apartment, written in Yiddish by a Mr. Lauber of New Haven, acknowledged receiving a 100-pound shipment of yeast from my grandfather and requested that he send an additional 150 pounds of yeast, which "should look good" and be "fresh." This postcard more or less substantiated my mom's recollection that my grandfather had sold a formula for yeast to raise the down payment to buy the Store. My grandfather did not sell a formula for yeast to

Fleischmann's. He was actually in the business of selling yeast. Perhaps my mom recalled the name Fleischmann's because for many years, from the end of the nineteenth to the early twentieth century, they had a retail store at the corner of Broadway and Ninth Street. Apparently, my grandfather's interstate transactions turned more on the issue of "fresh" than flesh.

My grandparents were well off, but they, unlike their sons, were not hoarders. They didn't meet the stereotype: they married, bought a house, managed to hold on to it throughout the Depression, had three children, enjoyed an occasional night at the Yiddish theater, and made sure that my mom and Uncle Harry went to college. Uncle Harry developed his hoarding skills on his own.

But those skills compelled him to create his unique filing system: keep everything. And that included the waiver from the Mann Act and the *Desiderata.* I have derived more meaning from the *Desiderata* than from the waiver. It turns out the *Desiderata* was written by a middle-aged Indiana attorney named Max Ehrmann. It seems Max always wanted to be a writer. In his early forties, he gave up his legal career, like me, and started to write full time. The *Desiderata* is his most famous work. It has inspired generations of readers with, at least from my perspective, its very simple conclusion: no matter how crazy your family may be, get over it. Only forgiveness can bring true understanding.

Yet, despite frequent readings of the *Desiderata,* my mind still wanders when I lie in the hammock in the back garden of my lake-front home, sipping a glass of Yarden Merlot, staring at the calm waters as the pine trees gently sway. Invariably, one question drifts in and drops anchor.

Did Max Ehrmann's mother ever tell *him* he didn't like money?

MORE ABOUT THE AUTHOR

MORE ABOUT THE BOOK

EXTRAS

More About the Author

ON IDENTITY AND THE WRITING OF *DOUGH*

We call ourselves what we are, but also what we wish to be.

—Leon Wieseltier

The story of the writing of *Dough* begins with a yellow number two pencil.

"Good writers can write about anything," my dad told me when I asked him for some direction after Mrs. Greenwald, my sixth grade teacher at PS 197 in Brooklyn, asked us to write an essay on a subject of our own choosing. "That pencil in your hand," dad added, "you should be able to write an essay about that pencil."

I've long forgotten what I wrote about, but it wasn't the pencil. I do recall feeling that being a writer was beyond my capabilities.

Yet, that is exactly what I wanted to be. But ever the dutiful son, at my parents' behest, I majored in accounting at Brooklyn College. As a result, for most of my adult life, when someone asked me what I did, I said I was a CPA. Although I managed to get some of my technical articles published in such literary hot-beds as *The CPA Journal* and *Taxation for Accountants*, that pencil in my hand got a lot of use preparing tax returns and not much else.

But after the surprising events described in *Dough*, I had a second chance. I gave up accounting and when someone asked me what I did, I said, "I'm an apprentice writer."

During my apprenticeship, unlike the sixth grade, I didn't have to think too hard for material. Life had presented me with a far more interesting story than a yellow number two pencil. But it took me years to bring the quality of my writing close to the level of the fascinating narrative that had been dropped into my lap.

Over the five years it took me to write *Dough*, I received many rejection letters. I suffered from what is known in the profession as premature submission. It reached the point where I avoided going through our daily mail. Instead, my son would look through the mail, recognize my now all too familiar self-addressed return envelope containing the bad news, and yell out to me, "Dad, you got another one."

The typical rejection letter from literary agents read, "This is a charming story, but it is also what editors call *a quiet book* and unfortunately, they don't want to buy quiet books."

Luckily, I had five literary angels, all gifted writers in their own right, whose suggestions sustained me. I learned the key is to have a pencil with a very good eraser.

Anne Neumann, my instructor at a writing workshop I took at the Princeton Arts Council, acted as the benevolent English professor I never had in college. She drilled me on the importance of sentence structure, comma use, and pacing. Kitsi Watterson, a writing professor at the University of Pennsylvania, nearly drowned getting through my first draft: seventy-five single-spaced pages that must have seemed like one long run-on sentence. Kitsi pointed out those riffs worth breaking out the life preservers. Esther Schor read a later draft of *Dough* with the same care and attention to detail she gives her very lucky students at Princeton University. Elizabeth

Frank, a Pulitzer Prize winner, and my teacher in a nonfiction workshop at the 92nd Street Y, taught me it's not enough to merely write, but to strive to "write well."

The incomparable Hettie Jones, whose memoir writing workshop I took at the 92nd Street Y, gave me much needed assistance in sequencing my manuscript. After hearing the story that was to become the chapter, "Awakening," Hettie strongly suggested that was where I should begin my story. When a writer who traveled in the same literary circles as James Baldwin, Jack Kerouac, and Allen Ginsberg tells you where to start your book, you listen.

That change set in motion my alternating chapters between the distant past and the near past to show how subsequent events can alter our memories. Once the book's pattern was set, I knew it worked. It was like hearing the tumblers of a safe click into place after finally getting the combination right.

In the same way my manuscript went through several drafts, it also went through several working titles. While cleaning up my uncles' apartment, we found a piece of paper stuck in an old bank book that had the words *non-routine transactions* stamped across it. Initially, that became my working title. Later, I used *The Boy Who Didn't Like Money.*

I never felt comfortable with those. *Nonroutine Transactions* was too generic and *The Boy Who Didn't Like Money* was too long. My wife, Nurit, suggested I call the book, *From Dough to Dough.* I liked that, but I took out my pencil and did some editing. In the end, all that remained was *Dough.* But in Nurit's honor, the first word of *Dough* is bread and the last, money.

Not everyone is fortunate enough to have a cast of characters quite like mine, but we all have a story to tell. Some of us are content keeping it to ourselves, or just sharing it with family or friends. When you write a memoir with the hope of publication, however, you must put your story through three sieves to see what, if anything, remains. First, would people who don't know you be interested in your story? Second, how much of your personal life are you willing to disclose? And last, what tone of voice are you utilizing?

In my case, my family story got through the first sieve with much to spare. But then I was unsure how much to reveal. The first essays I wrote for publication as stand-alone pieces, "Waiting for Cohn" and "The Food Stamp Seders," didn't reveal any financial details. Eventually, I worked up the courage to submit "Counselor Zachter" to the *New Jersey Lawyer Magazine.* Since the sky didn't fall down when it was published, I gained the confidence to submit the full manuscript for publication. After *Dough* was published, I started to tell people I felt like I was out of the closet.

"No, Mort," my friend, David Robinson, replied, "It's more like you stepped out of the vault."

Lastly, I gave careful consideration to my tone of voice. In my early drafts, it was angry. But by my final draft, I had matured both as a writer and as a person. I got my meaning across with less bitterness. The rule that allowed me to do this originates with one of my favorite writers, Calvin Trillin. He calls it the Dostoyevsky Rule: In nonfiction, never write anything terribly negative about anyone—unless you write as well as Dostoyevsky.

After *Dough* was published in 2007, I felt comfortable enough to say I was a writer. But as a promotion for a joint-reading I gave with another author that fall in downtown Manhattan, the following appeared on the Web site, Lucid Culture:

> *Thursday Oct. 4, 7PM New School professor and author Jocelyn Lieu reads from* What Isn't There *at Mo Pitkins. . . . Also on the bill: humorist Mort Zachter.*

During my reading, the audience laughed a few times. The fact that Mo Pitkins had a liquor license certainly helped.

I don't think I will ever call myself a humorist, but I'm not ruling anything out. After all, I can now tell my dad I've finally written an essay about a yellow number two pencil.

ABOUT THE SUCCESS OF *DOUGH*

The Making of a Surprise Hit

The line between failure and success is so fine that we scarcely know when we pass it— so fine that we often are on the line and do not know it.

—Ralph Waldo Emerson

Many years, and drafts, after I started to write *Dough,* I submitted my manuscript to a national writing contest. Months passed without a word. I finally gave up hope. My wife, Nurit, said it was time for me to view my writing as a hobby, not a vocation. She suggested I consider working as a reporter at the local newspaper. I placed my manuscript in the bottom drawer of my desk and prepared to move on with my life.

But at eight in the morning on Friday, May 26, 2006, our phone rang. A woman with a name straight out of a V. S. Naipaul novel, *Supriya,* said she was calling on behalf of a writing organization called the AWP (for more information on the Association of Writers & Writing Programs visit their Web site at www.awpwriter.org). It had been so long since I submitted my manuscript that I had no idea what she was talking about. I thought she was trying to sell me a subscription to a literary journal. Not quite. She said I had won the 2006 AWP Prize in nonfiction and my manuscript was going to be published by a university press. After years of writing and rewriting, of loneliness and frustration, I felt elated.

I soon heard from the acquisition editor, contracts manager, and marketing director at the University of Georgia Press. Their southern accents had me contemplating the sweet smell of pecan pie and wondering why the only Faulkner I had ever read was *The Unvanquished.* Getting published was top drawer. Nurit stopped talking about *my hobby* and emailed most of the English-speaking world that she was married to an award-winning writer.

That was the beginning of a surreal journey that would ultimately make me the University of Georgia's best known Garden State export since Heisman Trophy winner Herschel Walker played for the New Jersey Generals in the old United States Football League.

Surprisingly, the first inkling *Dough* was on the rise came from our nation's capital. Producers at both National Public Radio's *All Things Considered* and *All Things Considered Weekend* wanted to interview me for a national broadcast. *Weekend* won out. They sent their onetime middle-eastern correspondent, Jacki Lyden, to New York to interview me in the Ninth Street Bakery.

Yes, the Store still exists in the same spot my grandmother selected in 1926. A Ukrainian couple owns it now and the chocolate babka is still excellent. When you're in New York City, be sure to visit Ninth Street. Tell Oleg and Tetyana I sent you.

The interview with Jacki, a veteran reporter who had interviewed Arafat and King Hussein of Jordan was memorable.

At one point, she looked me straight in the eye and asked if I loved my uncles. That wasn't a question I had anticipated.

And my mom and dad's reaction to the publishing of *Dough* also surprised me. They had read the manuscript as I was writing it. At that time, only my mom had a comment. She didn't like my referring to a gray smock she wore in the Store as being *dirty*.

"No," she said to me after reading that line, "I never wore a dirty smock in my life." In the final draft of *Dough* that smock became *steel gray*.

Later, after reading the hardcover version of *Dough*, dad told me I had written a "good book." His only request was to have me clarify, in one specific line, that it was Uncle Harry who sat on millions, not he and my mom. In the paperback version, I've respected his wishes.

Mom and dad are overjoyed with the book's success. During our daily phone conversation they often ask where *Dough* ranks on Amazon. We even have the request down to a single word. Dad just asks, "Where?" and I know exactly what he means.

But the best was yet to come. The *Los Angeles Times* gave *Dough* a glowing review. And on Thanksgiving weekend, 2007, after years of reading the *New York Times Book Review*, I opened it up to see *Dough* favorably reviewed. For any author, especially a first-time author who used to be a CPA, this is the literary equivalent of winning the lottery. And a few weeks before Christmas, HarperCollins gave the University of Georgia Press a holiday present by purchasing the paperback rights to *Dough*.

Not wanting to leave my family needy at holiday time, Sony Pictures made an "inquiry" into the availability of film rights. Mere "inquiry" was enough for my family to begin dreaming of which film star would play their part. After much debate, Woody Allen emerged as the leading candidate to play Uncle Harry. Jerry Stiller and Anne Meara are the popular choice for my parents. Charlie, my quick-with-a-quip brother-in-law, believes his part should be played by Billy Crystal. No word yet from Billy.

Nurit suggests Scarlett Johansson portray her. "In that case," I told my beautiful wife of twenty-two years, "I'll play myself."

And these dreams are not merely limited to family. Our friend, Ed Soffen, a physician with theatrical aspirations, pines to play Uncle Harry. I don't know about that, but if I have any say in the matter, Ed will play the part of the New York City cabbie.

Friends in Israel even got into the act trying to think of a title for the movie in Hebrew. There is no direct Hebrew translation for the word "dough." As a take off on the blessing on bread, *hamotzee lechem meen ha'aretz,* our friends came up with, *hamotzee kesef meen halechem,* which means "he who makes (extracts) money from bread."

But the person with the clearest vision for *Dough* is my mother-in-law, Ellen. At the age of eighty, after bravely battling cancer for over three years, her goal is to live long enough to see the filming of *Dough.*

"I have to be around," she said, "Who else will make sure Julia Roberts gets me right?"

And after I gently explained that the chances of a movie version of *Dough* being made anytime soon were slim, my mother-in-law replied, "Mort, call Spielberg's mother. She'll get him to do it."

In this way, hope found fulfillment in a *wry* book about rye bread.

READING GROUP QUESTIONS

1. This book has a bifurcated structure that alternates between the distant and near past. Why do you think the author might have used this nontraditional structure for the book? Did you find this structure effective or distracting?

2. What were you thinking after reading the last line of the opening chapter, *in their entire lives, my uncles never baked a thing*?

3. Did the phone call from Uncle Harry's stockbroker, Mr. Geary, in "Awakening" add to your confusion? Did the truth surprise you?

4. Does the chapter, "Waiting for Cohn," give you a sense of the Store as a part of the fabric of New York City history? How are your family's history and memories interwoven with their locale, and how does this affect your sense of place?

5. Which chapter was your favorite? Why?

6. If you found out your family had a secret fortune but you had grown up sleeping in the dinette with your head next to the refrigerator, how would you feel? Would your reactions have been the same as or different than the author's?

7. *Dough* has been praised for its non-judgmental tone. Do you think the author would be so forgiving if he had not inherited the lion's share of his uncle's money?

8. Do you have any relatives whose mysterious ways remind you of Uncle Harry? Do you have any relatives with a touch of the hoarder mentality? If so, how do they compare to the author's family? Do you think you're in for any surprises when they pass away?

9. Aging, and the changes it brings, is an important topic in the book. How does your own experience with aging (your own aging, and that of your family and friends) compare with the author's?

10. What are some of your family's deeply ingrained feelings about money?

11. The author has said that *Dough* was written with as few adjectives and adverbs as possible. Why might he have made this stylistic choice? Did you find this style of writing appropriate for the story?

12. Some memoir writers will only include scenes in which they were present. This approach fails to document stories that have been passed down only through the oral tradition of storytelling. Do you think the author was right to document stories such as "Waiting for Cohn" or "A Tale of Urban Renewal" that he learned of, in large part, only through stories his family told him? How did these stories contribute to or detract from the larger story at hand?

13. Since publication, the author has received emails from all over the country from people who shopped at the Store and knew Helen, Joe, and Harry. Was there a legendary local retailer in your neighborhood when you were a child? What memories do you have of your visits there?

STRUNK & WHITE: MY NEW BEST FRIENDS

Or What Happened When *The Elements of Style* Met *Dough*

Early on in my writing apprenticeship, I bought the classic writer's reference guide, *The Elements of Style* by Strunk and White. After well over half a century, *Elements* is still considered the definitive book on all matters grammatical, logical, and stylistic when it comes to writing. As I worked on my manuscript, *Elements* sat on my desk, always within easy reach. I referred to it so many times its pages became frayed and worn. Nurit, my Yekke (a German-Jewish person who likes everything neat and proper) wife, couldn't stand the sight of it, and bought me a beautiful new hardcover version. As a former accountant who is never quite sure whether or not the modifiers in his sentences are dangling, it became a trusted ally on the road to publication. As a way of reminiscing about my journey, I visited my old friends, Mr. Strunk and

Mr. White, to review how some of their commandments ultimately applied to *Dough*.

1. "Omit needless words."

Strunk and White believe that "Vigorous writing is concise." I couldn't agree more. Unfortunately, one editor who read my manuscript sent me a lovely rejection letter stating it was too short for her to publish. After publication, I considered sending the editor who rejected *Dough* because of its brevity a copy of *Elements* with this specific commandment highlighted in green.

2. "Choose a suitable design and hold to it."

This is easier said than done. My friends say successful writing requires a clear perception of the ultimate shape your work will take. For the better part of five years, I had no clue how to sequence my story. In 2003, the first full version of *Dough* that I was not too embarrassed to show people emerged. I bombarded my wife's reading group with a manuscript divided into the four seasons of a life, all beginning with a separate scene in the clean up of my uncles' apartment. I'd flash back in time based upon an object we would find (e.g., a white soup pot with a thin blue rim). It didn't work. The next version of *Dough* was strictly chronological; readers fell asleep before they discovered the double meaning of the title. Then after I took the memoir writing workshop with Hettie Jones, the riff that would ultimately become "Awakening" became my starting point. That third draft is the one that was ultimately published.

3. "Revise and rewrite."

If writing workshop instructors, literary agents, and editors all tell you that your writing needs work, and one even suggests you take your manuscript, put it in a box, and place it in a bottom drawer and forget about it, this is an easy rule to follow.

4. "Do not affect a breezy manner."

According to my buddies, affecting a breezy manner is "often the work of an egomaniac." We're now well into the era of the literary memoir, so I don't worry about this one much. Thanks to reality TV, Donald Trump, and Paris Hilton, I'm convinced I'm not an egomaniac—I'm a memoirist.

5. "Avoid the use of qualifiers."

The boys go wild on this one. "Rather, very, little, pretty—these are the leeches that infest the pond of prose, sucking the blood of words." I now usually avoid those words. But on the days I do use one, I make sure I'm home before nightfall and the front door is securely locked.

6. "Do not explain too much."

Despite the clarity of this wonderful five-word commandment, my good friends felt compelled to add, "Inexperienced writers not only overwork their adverbs but load their attributives with explanatory verbs." I assure you I never overworked an adverb. And I'm not strong enough to lift an overloaded attributive. In fact, I'm not even sure I know what an attributive is.

7. "Avoid fancy words."

I promise not to use the word penumbra in my next book. On the other hand, I'm still convinced Uncle Harry spent a lot of time in the penumbra of New York City.

8. "Do not use dialect unless your ear is good."

My mom always told me I was a good listener. I took this to mean I had not one, but two good ears. Therefore, you will find quite a bit of dialogue in *Dough*. If you enjoy the Brooklyn accent of the clerk in the Kings County small-estate department or Miss Essie's southern drawl, thank my mom.

9. "Be clear."

Coming from a family that always answered a question with another question, this one wasn't easy to follow, but I tried.

10. "Prefer the standard to the offbeat."

With the offbeat set of characters I had to work with, I never needed to turn to the "eccentricities in language."

Despite these good intentions, in the end, readers will decide if *Dough* ennobles Strunk and White's literary classic or if it is just another infestation in the "pond of prose." At a minimum, my one-word title omits needless words.

CUSTOMER FEEDBACK

Since the publication of *Dough*, I've been fortunate enough to speak with, and receive emails from, many former customers of the Ninth Street Bakery. In the fall of 2007, I gave a reading at the Mulberry Street Branch of the New York City Public Library and several former customers of the Store attended. At one point, I passed the microphone around and let them share their recollections of my uncles. It was a memorable evening.

Of all the emails I've received from former customers, the one that best captures the flavor of that evening came from Mr. John Feidelson.

> Mr. Zachter,
>
> While painting my front door yesterday I was, as I often do, listening to NPR. I nearly fell off my ladder when I heard your interview. I lived in NYC from 1977–92—most of that time on East 10th between A and B by Tompkins Park. Now, alas, I am in Brookline outside of Boston. During the early part of those years (before marriage) I lived on a steady diet of sandwiches. The bread was from your uncles' bakery, the salami and mustard from Kurowycky's (which I gather closed in June). I always regarded your uncles with a fascination, and to this day I have a strong memory of their shop: The decor was 1930s with a splash of '60s bright orange; the painted wood drawers barely containing a mismatch of plastic bags looked as though they might fall out at any moment; the

cracked tile floor defied pattern. I always went for the dark pumpernickel. Harry would say, "It's a nice bread." My wife and I often repeated that to one another with warm amusement. We called your uncles "The Bread Boys."

I never discussed much with Harry, but I remember the ritual of asking him, "Could you please slice the bread." The one technological marvel in the place—the bread slicer—gave variety to the transaction (to slice or not to slice?). Even as he pushed down on the cutter to speed up the process (or maybe to force a cut), it appeared that this was his moment to reflect. He stood there chewing a heel of bread, with long hair more yellow than white, and looking beyond me waited for the machine to stop. Once the clatter subsided he would stir from whatever respite he had experienced. This signaled the grab-a-plastic-bag-from-the-falling-out-drawer phase of our transaction. He'd scoop up the bread and with a flourish twist the tie around the bag. We'd exchange cash and change. The register was nearly as antique as your uncles but apparently it was a deep coffer.

A friend of mine remembered this story:

A customer was telling Harry that he should hire some young help so that he could move more "merchandise." Harry replied, "They're selling aspirin around the corner. Two bottles for a dollar. It pays to get a headache."

Thanks for the memories.

John Feidelson
Brookline, MA

THE LOWER EAST SIDE: THEN AND NOW

Change is inevitable, except from a vending machine.

—Anonymous

Where derelicts once roomed in flop houses, a luxury hotel on the Bowery provides turndown service for guests who sleep on 400 thread count Egyptian cotton linens. Nearby, the architecturally provocative seven-story New Museum of Contemporary Art towers over the Bowery.

The Lower East Side constantly renews itself. Whether this is good or bad depends on your perspective. For example, in the fall of 2006, for the first time in over a century, no Yom Kippur services were conducted at the First Roumanian–American Synagogue. Earlier that year, the 150-year-old Romanesque Revival building had been razed. When I walked past the site in the late summer of 2007, all that remained was a multimillion dollar real estate opportunity masquerading as a vacant, weed-strewn lot. Most likely, a residential condo will be built someday. Will its owners have any idea of what once stood there? Will they even care?

I remember attending services inside a sanctuary that could seat over 1,600 congregants. There I heard the last echoes of what was once called the "Cantor's Carnegie Hall," where Jacob Pincus Perelmuth (better known as Jan Peerce) once sang.

On the south side of Rivington Street change had already come. In the place where Mr. Lipshitz once sold *siddurs*, you can now buy cappuccinos.

A few blocks away, the Eldridge Street Synagogue survives. Why this synagogue was renovated, and the First Roumanian torn down, is a question for the rabbis and historians. While they debate, on Houston Street, customers still line up to buy lox at Russ and Daughters and potato knishes at Yonah Schimmels.

And despite all these changes, at 350 East Ninth Street, beneath a framed black-and-white photograph of Uncle Harry and Uncle Joe, the merchandise is still moving.

ACKNOWLEDGMENTS

I am especially grateful to my wife, Nurit, for her love, patience, and support.

Thanks to Elizabeth Frank, Hettie Jones, Anne Neumann, Esther Schor, and Kathryn Watterson for all their sage advice.

For taking the time to read, in whole or in part, early, half-baked versions of *Dough*, thanks to Nancy and Howard Alter, Mary Ciuffitelli, Rebekah and Andy Costin, Pam and Dean Edelman, Kate Fridkis, Steve Guggenheim, Vanessa Druscha Guida, Angela Himsel, Marty Katz, Michael Lapp, Steve Lawrence, Beth Morgan, Dottie Myers, Debbie Nathan, Ruth Ramsey, Jackie Rea, Fay Reiter, David Robinson, Cecilia and Irwin Rosenblum, Sandy Shapiro, Kathryn Silverstein, Sanjna Singh, Debbie and Ed Soffen, Albert Stark, Alison Stateman, Irene Waring, Bob Weinstein, Shoichi Yoshikawa, and all the members of Nurit's Princeton reading group and Elizabeth's nonfiction workshop at the 92nd Street Y.

Many thanks to my editors, Andrew Berzanskis and Jennifer Reichlin, and the entire staff at the University of Georgia Press, as well as the Association of Writers and Writing Programs.

And thanks to Kyoko Mori, who, for reasons that continue to astound me, selected *Dough* for the 2006 AWP Award for Creative Nonfiction.

And lastly, for turning me into a paperback writer, Carol Mann, Ethan Friedman, Jean Marie Kelly, Shelby Meizlik, Matthew Inman, and Kimberly Chocolaad.

AUTHOR'S NOTE

The December 27, 1947, and December 28, 1947, editions of the *New York Times* are the primary sources for the chapter "Waiting for Cohn."